George Miller Beard

Stimulants and Narcotics

George Miller Beard

Stimulants and Narcotics

ISBN/EAN: 9783337236489

Printed in Europe, USA, Canada, Australia, Japan

Cover: Foto ©berggeist007 / pixelio.de

More available books at **www.hansebooks.com**

PUTNAM'S HANDY BOOK SERIES

CHAS. JOSSELYN.

STIMULANTS AND NARCOTICS;

MEDICALLY, PHILOSOPHICALLY, AND MORALLY
CONSIDERED.

BY

GEORGE M. BEARD, M.D.

NEW YORK:

G. P. PUTNAM & SONS.

1871.

PREFACE

THE objects of this work are to give a brief description
and history of the principal Stimulants and Narcotics ; to
discuss in a manner at once scientific and popular their
good and evil effects, as modified by race, climate, age, sex,
individual temperament, habit, the state of health, and the
progress of civilization, and to indicate the methods by
which the enormous evils to which they give rise may be
successfully opposed.

So far as I know, this is the first systematic attempt of the
kind that has ever been made. Although the literature of the
subject is very extensive, yet it mostly appears in the form
of special pleas, either for or against some one of the more
prominent varieties of stimulants and narcotics, as opium, or
tobacco, or alcoholic liquors. Works written in a spirit of
partisanship can never satisfy the honest lover of truth.

If the unaided emotions could have determined this great
question ; if the wish had always been father to the truth,
as well as to the thought ; if our philanthropists and legis-
lators had been trained to habits of scientific reasoning, any
further discussion of the subject would be superfluous.

As it is, there is no other social question concerning which
so much has been stated, and so little known ; upon which
men have felt so strongly and reasoned so feebly. While very
many of the choicest spirits of the world have thought
intensely on this subject; have spoken and written eloquently

upon it ; have quarreled over it, wept over and prayed over it, comparatively few have reasoned upon it, and those few have been too meagerly supplied with facts to arrive at any just conclusion. Even some of the men of science who have entered upon the study of this subject have displayed a spirit the reverse of scientific.

This work is now prepared in the belief that there is a large and increasing number in this country who have not been wholly satisfied with the manner in which the subject has thus far been treated ; who perhaps suspect that something more than witty stories, or stirring songs, or eloquent exhortations are needed to solve a great social problem ; who are beginning to feel the propriety and the necessity of discussing a question that is primarily scientific by scientific methods of reasoning ; who in the spirit of honest inquiry are eagerly seeking for information to aid them in forming their opinions ; and who are ready and willing to attend to facts that are novel, and to listen to views that are directly opposed to their own, and who recognize the need of patience and wisdom as well as of courage and enthusiasm in conducting a great reform.

The facts contained in this work have been gathered with not a little labor, and many of them are now for the first time presented in an accessible form ; and the method of treating the subject throughout is radically different from that to which the popular mind has been accustomed.

I desire to call special attention to the comparison made between the climate of Europe and North America, which has so important a bearing on this question ; to the statistics showing the relation of ignorance to intemperance and other vices ; to the sketch of the drinking customs of different countries, and to the history of temperance legislation from the earliest periods down to the present time.

G. M. B.

New York, *September 1st,* 1871.

SOURCES OF INFORMATION IN REGARD TO STIMULANTS AND NARCOTICS.

ALTHOUGH so much has been written on stimulants and narcotics, yet the number of really valuable works on the subject is quite limited. Strange to say, some of the most important information on the subject is found in works in which stimulants and narcotics are only treated of incidentally.

Authors who have opinions on the subject are painfully numerous ; authors who have facts that may aid us in forming our opinions are easily counted.

The following are some of the most valuable and accessible sources of information on all the branches of the subject discussed in this work, and in which many of the facts I have recorded can be verified.

To have encumbered the body of the work with copious footnotes and references, would have repelled the large body of readers for whom it was designed, without adding materially to its scientific value.

A fair résumé of the leading medical and scientific researches and opinions may be found in the following writings :

FRANCIS E. ANSTIE, M. D.—Stimulants and Narcotics, 1864.
——— The Use of Wines in Health and Disease, 1871. Republished from the Practitioner.

JOHN HARLEY, M. D.—The Old Vegetable Narcotics, London, 1869.

EDMUND A. PARKES, M. D.—A Manual of Practical Hygiene, London, 1866.

H. LETHEBY, M. D.—On Food, London, 1870.

LALLEMAND, DUROY AND PERRIN.—Du Rôle de l'Alcool et des Anesthésiques Paris, 1860, and Union Médicale, 24 Décembre, 1863.

BAUDOT.—De la Destruction de l'Alcool dans l'Organisme. Union Médicale, Nov. and Déc., 1863.

DUCHEK.—Ueber das Verhalten des Alkohols in thierischer Organismus. Viertel Jahrschrift für die praktische Heilkunde. Prague, 1833.

MAGNUS HUSS.—Chronische Alkohols-Krankheit, 1852.

W. MARCET, M. D.—On Chronic Alcoholic Intoxication.

W. A. HAMMOND, M. D.—A Treatise on Hygiene, 1863.

EDWARD SMITH, M. D.—On the Action of Alcohol, 1861.

E. A. PARKES and WOLLOWICZ.—Experiments on the Effects of Alcohol on the Human Body. Druggists' Circular, Nov., 1870.

Among the representatives of extreme views in the profession that are now pretty well exploded are :

W. B. CARPENTER, M. D.—Use and Abuse of Alcoholic Liquors in Health and Disease. London, 1851.

T. FORBES, M. D.—Physiological Effects of Alcoholic Drinks. 1848. Massachusetts Temperance Society.

Among the authorities on the description and history of stimulants and narcotics are :

GEO. H. LEWES.—Physiology of Common Life, 1860.

JAMES F. JOHNSTON.—The Chemistry of Common Life, 1856.

JAMES F. CLARKE.—Ten Great Religions, 1871.

W. E. H. LECKY.—History of Rationalism in Europe, 1868, vol. ii., pp. 321, 322.

Mr. Lecky refers to the following works on the subject :

D'AUSSY.—Hist. de la Vie Privée des Français, (Paris, 1815,) tom. iii. pp. 116-129.

PIERRE LACROIX.—Histoire des Anciennes Corporations, p. 76.

PELLETIER.—Le Thé et le Café.

CABANIS.—Rapports du Physique et du Moral, 8vo.

McPHERSON.—Memoirs, Annals of Commerce, vol. ii. pp. 447-489.

Other works on the same subject.

VON BIBRA.—Die Narkotischen Genussmittel und der Mensch. Nürnberg, 1855.

JAMES RICHMOND SHEEN.—Wines and other Fermented Liquors, London.

Robert Tomes.—The Champagne Country.

Jacob Biglow, M. D.—Nature in Disease, 1859.

Alonzo Calkins, M. D.—Opium and the Opium Appetite, 1871.

John P. Hedges.—History of the Excise Law, 1856.

C. Darwin.—Descent of Man, 1871.

Some works of travel give valuable information on the subject.

Madden.—Travels in Turkey.

Wilkes.—United States Exploring Expedition.

Pöppig.—Reise in Chile, Peru und auf dem Amazon Ström.

Von Tschudi.—Travels in Peru.

Francis Galton.—Tropical South Africa.

Weddell.—Voyage dans le nord de la Bolivie.

Hooker.—Himalayan Journals.

Meyers.—Life and Nature under the Tropics.

Fortune.—Tea Districts of China.

Thomas W. Knox.—Overland through Asia, 1870.

Samuel Hazard.—Cuba with Pen and Pencil, 1871.

Richard J. Bush.—Reindeer, Dogs and Snow Shoes, 1871.

Paul B. Du Chaillu.—Explorations and Adventures in Equatorial Africa, 1862.

——————————— A Journey to Ashango Land, 1867.

For information on the comparative influence of European and American climate, and for various statistics on the different themes discussed, I may refer to the following :

M. E. Deson.—The climate of the United States, and its Influence on Habits of Life and Moral Qualities. Bost. Medical and Surgical Journal, March 16, 1871, p. 175.

W, Wislizenus, M. D.—On Atmospheric Electricity. Transactions of St. Louis Academy of Medicine.

Geo. M. Beard, M. D. and A. D. Rockwell, M. D.—Medical and Surgical Electricity, pp. 93-95 and 288-290.

S. W. Mitchell, M. D.—Wear and Tear, 1871.

Stadtisches Jahrbuch für Volkswirthschaft und Statistik. Berlin, 1869.

Reports on the Subject of a License Law, by a Joint Special Committee of the Legislature of Massachusetts, together with a Stenographic Report of the Testimony taken before said Committee.

Second Annual Report of the State Board of Health of Massachusetts, 1871.*
Report of the Commissioners of Education, 1870.
Police Reports of Liverpool, London and New York.

Besides these authorities, I am under obligations to a large
number of individuals in various countries, who, in many ways
have aided me in prosecuting my inquiries. My thanks are especially due to Meredith Reed, Jr., our Consul General at Paris ;
Consul General Morse at London ; Mr. Farr, Register General
of England ; Hon. David A. Wells, Ex-Commissioner of Internal
Revenue ; Hon. Edward Young, of the Bureau of Statistics ; Dr.
George Derby, and Dr. Henry I. Bowditch of Boston, to Rev.
James Beecher, and to Messrs. Beebe and Co., the well-known tea
merchants of New York.

---

* The State Board of Health of Massachusetts addressed a circular of
inquiry concerning the use and effects of alcoholic liquors " to the American
Ministers at foreign courts, and to the Consuls of all the principal ports of the
globe." The replies, which are numerous and very interesting, have been of
great service to me, especially in the preparation of the third chapter.

TABLE OF CONTENTS.

INTRODUCTION.

CHAPTER I.

CHAPTER II.

CONTENTS.

CHAPTER III.

THE PHILOSOPHY OF INTEMPERANCE, AND THE PRINCIPLES BY
WHICH IT SHOULD BE TREATED ....

PAGE

CHAPTER IV.

STIMULANTS AND NARCOTICS, IN THEIR MORAL, SOCIAL AND
ECONOMIC RELATIONS, WITH PRACTICAL SUGGESTIONS CON-

xiv CONTENTS.

STIMULANTS AND NARCOTICS.

INTRODUCTION.

DIFFICULTIES AND COMPLICATIONS OF THE SUBJECT.

THE subject of stimulants and narcotics has become one of the most difficult social questions of our time.

The reasons why it is thus difficult are these :

1. It is a subject of vast complications.

Ten thousand influences are continually acting upon us for good or ill, and hard enough it is, even for the scientific and impartial observer, to tell which contributes the most to make us what we are. The human system is a vast ocean into which myriads of streams run, and no one can say just how much each stream brings with its tide. Questions of race, of climate, of education, of religion, of legislation, of morals—all are so woven into this great question as to weave a vari-colored pattern ; and to pick out each thread one by one is a task at once painful and patience-demanding, and few there are who are ever measurably equal to it, and so, like angry schoolgirls over a skein of tangled silk, we pull and tear it in all directions, hoping, by some lucky chance, to untie the knot.

2. It is related to many difficult departments on

which people have very strong feelings, and therefore it can be discussed calmly and rationally only by those who have well-trained minds.

It is so closely connected with morals and religion and politics—subjects on which men feel much more than they reason—that as soon as they attempt to find out the truth in regard to stimulants and narcotics, the emotions rise up like a flood and, so to speak, drown reason out.

Now while the emotions are often indispensable to urge and drive on the reason, yet at the last moment, when the mind is to be made up, and a decision is to be finally given—when we enter into truth,—reason should detach itself from the emotions and go alone on its course, just as in entering a railway terminus the train is detached from the engine and switched off to its own track.

3. Another reason why this subject is so difficult, is that it has only been recently studied.

Time is a great solver of life's problems, and we have not had time to solve this question, for we are just beginning to open our eyes to it.

Men have probably always been intemperate, more or less, but only within the past half century have they systematically reasoned about intemperance.

Partly on account of the great increase of intemperance in our modern civilization among certain races, and in certain countries, the reasons for which I shall endeavor to explain, and partly on account of the improved moral tone that has accompanied our civilization, Great Britain, and America particularly, have during the last forty or fifty years been agitated, and almost convulsed on this subject of temperance.

So suddenly have the use and the abuse of stimulants and narcotics increased in modern times, that we are just now opening our eyes to their nature, their history, and their effects.

Those whose attention has been most earnestly called to the subject, and whose sympathies have been most warmly enlisted in it, have not yet sufficiently recovered from the shock caused by the discovery of the evils of intemperance, to consider the broad subject of stimulants and narcotics in a calm, candid, and scientific manner.

As an army marching unsuspiciously on its way, when, unexpectedly, it encounters an enemy in ambush, is at first thrown into disorder and confusion; as a merchantman, sailing under gentle breezes, when it is suddenly struck by a squall is at once thrown on its beam ends, and put in peril of utter disablement until the commander has had time to assert his authority, and each man has regained his appointed place; so our civilization has so suddenly encountered the full fury of the blast of intemperance, that philanthropists and philosophers have been at first so overwhelmed that they could not command the presence of mind necessary to devise systematic measures of relief; and in their haste and confusion and terror have rushed to wild extremes, have eagerly sought the aid of the noisiest charlatans, and in their despair have tried in succession every remedy that was loudly advertised, only to be disappointed by all; and even now are but just beginning to recover their breath and look the evil bravely in the face, and to rally their undisciplined and demoralized forces under the leadership of reason.

For the **sake of** convenience, and in order to keep each branch **of the** subject measurably distinct, I shall divide **this treatise** into four sections, as follows :

1. *The Definition, Description* **and** *History of Stimulants and* **Narcotics.**

2. *Their effects* **as** *modified by race, climate, age, sex, temperament, habit, state of health, adulteration, etc.*

3. **The** *Philosophy of Intemperance, and the principles by which it should* **be** *treated.*

4. *Stimulants and Narcotics in their moral, social, and* **economic** *relations,* **with** *practical suggestions concerning their use.*

THE DEFINITION, DESCRIPTION, AND HISTORY OF STIMULANTS AND NARCOTICS.

THE popular impression is that a stimulant is something that exalts and enlivens and whips up the powers, while a narcotic is something that produces sleep or stupor.

It is also a popular belief that stimulation must be followed by a corresponding reaction, and that it imparts no strength to the system, but merely calls into action latent and unused forces.

Stimulants, when given in stimulating doses, are not followed by any reaction, and they do add to and increase the powers of endurance.

All definitions and classifications of medical or alimentary substances must be more or less unsatisfactory.

It is impossible, I may say at the outset, to give any definition of stimulants and narcotics that can be universally acceptable. I here adopt the leading distinction drawn by Anstic, using mainly my own phraseology.

Stimulants are those agents which correct, economize, or intensify the forces of the system. Narcotics are those agents which produce a greater **or** *less degree of paralysis of some portion of* **the** *nervous* **system.**

The signs **of** stimulation are relief **of** fatigue, irritation, and pain ; equalization **of the circulation ;** improvement in sleep and in nutrition, and increased capacity for mental and manual toil.

The signs of narcotism are, in the first stage, flushing of the face, dilatation of the pupil, mental disturbance of **various kinds, as** evinced by the exhibition of garrulity, ugliness, **and so** forth, nausea, tremor, spasms, **convulsions,** and other evidences of lack **of** co-ordination ; in the last stages, delirium, stupor, **stertorous** breathing, **and** death.

These **very opposite** effects, **stimulation and** narcosis, **are produced by** the same agents, being modified by the **dose, the age, the sex, the temperament, the** condition **of health.**

The general law is that small **doses produce** stimulation, and large **doses narcotism ;** but small **and large** are relative terms, **or** what would **stimulate one may** narcotize another. What would stimulate at **one** time **of life,** at another narcotizes. (The quantity that may **narcotize the majority** of women **and** children, may only stimulate the **majority** of men.)

This distinction **between** stimulation and narcosis is radical and **important, for stimulation** is **a** beneficial **process, while** narcotism, when frequently repeated, and long continued, **can** never be otherwise than injurious.

The **method of** distinguishing these two orders of effects **will be** described further **on,** and in greater detail.

The stimulants and **narcotics that are in** habitual use

among men may nearly all be embraced under three grand divisions :

1. *Those which contain* alcohol—*fermented and distilled liquors.*

2. *Plants and vegetables that are used in substance, or in infusion or decoction,* by chewing, smoking, snuffing, or by *injection.*

3. *Volatile preparations that are used by inhalation, or by internal or external administration.*

Under the *first* division are included all liquors, fermented or distilled, that contain alcohol in any quantity.

Under the *second* division are included tea, coffee, cocoa, chocolate, opium, tobacco, haschish, Siberian fungus, etc.

Under the *third* division are included æther, chloroform, etc.

FIRST DIVISION OF STIMULANTS AND NARCOTICS.

FERMENTATION.—Fermentation is a kind of decomposition that vegetable substances undergo when placed in contact with air and moisture. The starch is converted into sugar, and the sugar into alcohol and carbonic acid. This is called vinous fermentation.

The liquid thus obtained always contains more or less alcohol, and receives different names according to the substance from which it is obtained.

Wine is the fermented product of the juice of grapes, currants, gooseberries, blackberries, elderberries, etc.

DISTILLATION.—When fermented liquors of any kind are boiled over a fire, the alcohol rises in the form of vapor, and when conducted through a coiled receiver it condenses into a liquid condition. **This** process **is** called distillation.

The product of the distillation **of wine is** called brandy or cognac ; of molasses, **rum** ; of corn, potatoes, rye, or barley, **malt** liquors, brandy or whiskey. When juniper berries are added before distillation **the product is called** gin.

Although the Chinese probably discovered the process of distillation thousands of years before **it was** known in Europe, they yet kept it, like all **their other** discoveries, entirely to themselves.

The Greeks and Romans were entirely ignorant of the art, and not before the 7th century have we any mention of the use of the still. At first fermented liquors were distilled, **but in the** 12th century Albucasis taught the art of distilling wine.

The first author who speaks distinctly of the distillation of wine is Arnauld de Villeneuve, a physician of the 13th century.

Moon Plant (*Soma, Haoma,* or *Asclepias acida*).—Comparative Theology carries **us back to a** time that probably long antedated the culture of the vine. Both the Zend Avesta, the sacred writings of Zoroaster, and the Vedas, the sacred writings of the Brahmins, speak of a sacred plant, the fermented juice of which was employed **in sacred rites.** This sacrament began in India far back in the misty twilight of the Vedic age, and **even** yet is occasionally observed. At this sacrament hymns were sung which have come down to **the** present day. It was declared that sacrifice with this **sacred** plant expiated sin, and the Brahmin was forbidden to taste any other intoxicating beverage.

The taste of the Moon-plant is described as astringent, bitter, very disagreeable, and it is said to be quite intoxicating.

In Persia we find that an important part of their sacred writings are devoted to this Moon-plant worship, and to the hymns that accompanied the ceremonies. At that time it was probably the only intoxicating beverage that was known, and its effects were so delightful and so peculiar that it was believed to be the gift of the gods, and hence was offered to them with sacred ceremonies.

It was believed that wisdom came from it, and health, and increased strength of mind and muscle, victory in battle, talented offspring, and length of days.

This identity of the sacred rites of the Persians, and Indians, is one among many other evidences that they both came from the same Aryan stock ; and hence we know that in still more distant ages the use of this intoxicating plant derived its origin. What Bacchus was to the ancient Greeks, such, thousands of years before, was this Moon-plant to the early Aryans.

GRAPE WINE.—We learn from the Egyptian hieroglyphics that wine was in use in Egypt thousands of years ago, and that the effects of intoxication were also not unfamiliar.

It is claimed that the vine originated in India, and thence spread to Asia Minor, Africa, Greece, and Europe. Originally the wine was very simple, merely the expressed juice of the grape. The effect of age on wine was early noticed, for Homer speaks of wine in the 11th year, and Pliny states that he had drunk wine two hundred years old.

ORIGIN OF **THE** USE OF WINE.—There is a very queer myth that a lady of the harem of King Jemsheed of Persia, becoming tired of life and desiring to commit suicide, stole some of the fermented juice of the grape, which the King had had locked away for security; supposing it to be poisonous, she drank largely of the strange fluid, (for up to that time grapes were raised for eating only,) became drunk, recovered, and being delighted with the sensations of intoxication, drank again until all was gone.

The King took the suggestion and popularized wine.

The vine is said to have been introduced into England by the Romans.

HOPS.—The hop has been known in Germany for ages, and by the Germans was first used in malt liquors. Hop gardens are mentioned as early as the 9th century, and in the breweries of the Netherlands, hops were used in the early part of the 14th century, and were introduced into England in the early part of the 16th century. The hop is of value in the beer in several ways. It arrests fermentation and so prevents souring. It gives it a pleasant bitter taste, and it adds to its stimulating or narcotic effects.

ALE or BEER is obtained from a decoction of *malt* and *hops*. The malt for beer is prepared by grinding, mashing, boiling, cooling, fermenting, vatting and clearing the grain. The modern improvements in brewing require eight processes. There are three kinds of malt : pale or amber brown, and roasted or black.

PORTER or STOUT is made from roasted or black malt.

The fermented product of the juice of apples is *cider ;* of the juice of pears, *perry.*

There are a number of substitutes for the hop in beer, some of which are quite injurious.

The chief of these is *Cocculus Indicus,* or *Levant nut.* It is the fruit or berry of a beautiful plant that grows in the Indian Archipelago. It gives a bitter and a rich taste to beer, makes its color darker and its effect far more intoxicating. The beer of the poorer classes is largely drugged with this powerful agent.

The *mead* of the Abyssinians is mingled with a bitter bark called "hectoo," with the leaves of a tree called "keesho" and with a root called "taddo."

Little is known of the chemical character of these substances, but they are believed to be all possessed of more or less stimulating or narcotizing properties.

Another substitute for the hop in Sweden and North Germany, is "marsh ledum," or "wild rosemary." Its effects are somewhat like those of *Cocculus Indicus.*

The English adulterate their beer with "clary" and "saffron," both of which substances decidedly affect the nervous system.

In North America "Labrador Tea" is sometimes used to adulterate beer, as a substitute for hops.

VARIETIES OF ALCOHOLIC LIQUORS USED THROUGHOUT THE WORLD.

The varieties of fermented and distilled liquors now in use throughout the world may be numbered by thousands. In Hungary alone there are four hundred kinds of wines ; and France, it is said, produces between one and two thousand varieties. Alcoholic liquors are not only made from different substances, but by different

modes of preparation, and are known in different countries by different names.

Thus, in South America the natives use "*cocoanut wine*,"
which they manufacture by the very simple process of
burying the cocoanuts in the sand by the shore, near the
edge of the tide, and leaving them there until the milk
ferments.

The peasantry of Switzerland and some parts of Germany rejoice in "schnapps," a species of brandy distilled from potatoes. Modern Greece depends on
"rakée," made of the lees of wine and figs; and Constantinople drinks "mastica," a rum flavored with mastica and brandy. The "whiskey of Spain" is "aguardiente."

The peasantry of the Rhine provinces call their
cheapest wine, on which many of the quite poor depend,
by the hard name "tietz." "Toddy," a word long ago
Anglicized, seems to have come from Ceylon, where it is
applied to a wine made of thè palm.

Wine made from the sugar cane, the West Indians
call "tafi," and the Brazilians "cachaça;" and by the
natives of Guarapo, sugar cane wine is "guarapo."

The Hindoos use "arrack," made from rice, which is
their great dependence. The alcoholic liquor made
from rice receives several names, according to the locality.
The Chinese call it "samshoo;" the Japanese "sacio"
or "saki."

The Indians of Florida prefer a "black drink," of
the leaves of the "emetic holly." A drink of the Abyssinians is "Tallab," and "vodki," the "whiskey of Russia" is distilled from the potato.

The Indians of the Andes indulge in a strong drink
made of the thorn apple.

The fermented drink made from the millet the Abyssinians call "bouza," and the Crim-Tartars "mur-wa" or "millet beer." The "toddy" of Sumatra is called "neva."

Several varieties of palm produce wine, and it is said to be used by a larger number of the human race in Africa, India, South America, and Oceanic Islands, than the juice of the grape. One palm tree yields from two to six pints of sap, which in a few hours ferments, and when distilled makes a powerful brandy. The natives of Northern Africa drink the wine of the date palm ; to this they give the name "lagmi." The Africans indeed use scarcely any other alcoholic liquor.

Of all the methods of causing fermentation, that of the South American Indians, in the valley of the Sierra, is the most remarkable. The maize malt—which is pre-pared by moistening Indian corn until it sprouts—is *chewed* by all the members of the family, and any stran-gers that happen to be present, all seated around in a circle.

As fast as each one thoroughly chews his handful, he throws it into a large calabash standing in the center. When the whole heap of corn is by this strange pro-cess reduced to a pulp, it is mingled with hot water and allowed to ferment.

The advantage of this process is, that the saliva changes the starch of the corn into sugar, which after-ward ferments. Beer made in this way is highly prized, and is offered to guests as a special compliment.

Beer thus prepared is called "chica" or "maize beer." "Chica" can be made also from grapes, pine-apples, rice, barley, peas and bread.

The Russians make a beer by mixing rye flour, and

barley flour, and water, and allowing the mixture to ferment. This they call "quass," or "rye beer." It is a "sour, thick drink," like the millet beer of the Africans. Every tyro in geography knows that the Tartars make a drink of mare's milk by adding yeast. The advantage of mare's milk is, that it contains more sugar than the milk of the cow. The fermented drink made in this way, the Tartars call "koumiss," or "milk beer."

When the "koumiss" is distilled, a milk brandy is produced, which the natives call "arraca."

In the islands of the Pacific "kawa," or "ava" is prepared from long pepper, in the same way as the chica of South America, by chewing. Like the Saoma of ancient days, it is used in religious ceremonies. The drinking of the ava is the first duty of the day, just after the morning prayer.

The favorite drink of the Mexican is "pulque," or "octli," or "agave wine," which is produced by fermenting the sap of the American aloe. This has an acid resembling that in cider, and a very disagreeable odor, but the taste is cool and refreshing. The brandy distilled from this is called "aguardiente," or "Mexical."

The Arabians prepare a fermented drink from milk, which they call "leban," and the Turks have a similar drink called "yaourt."

SECOND DIVISION OF STIMULANTS AND NARCOTICS.

TEA.—The tea-plant is a small shrub, growing to the height of seven or eight feet. It is chiefly raised in the valleys and on the sloping hillsides of China and Japan.

The first use of tea as of wine, dates from a mythical story. Darma, an Indian prince, visited China about the year of our Lord 510 on a mission of religion. In order to obtain greater influence over the people, he devoted nights and days to prayer. After several years, he fell asleep, overcome by fatigue ; when he awoke he cut off his eyelids and threw them on the ground, that he might the better remember to preserve his vow of vigilance. The following day he returned to the spot, and found that the eyelids had become a shrub, before unknown. Some of the leaves of this he ate, and they quickened his spirits and restored his vigor. He recommended the plant to his disciples, and it became popular.

It is tolerably sure that tea did not come into general use in China until about 1200 years ago, and in Japan two centuries later.

It was introduced into England in the latter part of the 17th century.

In the diary of Mr. Pepys, Secretary to the Admiralty, 1666, is found this record : "I sent for a cup of tea, a Chinese drink of which I had never drunk before."

In 1678 an important work on tea appeared. The writer lauded the beverage as "the infallible cause of health" and the cure of every ill. He thought that *two hundred cups daily* would not be excess. Poets and philosophers joined in praise of the new drink, and the East India Company, by whom it had been introduced, found it a source of increasing profit.

But the sentiment was not all one way. Against the introduction of tea were arrayed the mighty names of Boerhaave and Van Swieten, but in vain. Like coffee, like tobacco, like alcohol, it has prospered most when it has been most opposed.

At the present time, 3,000,000,000 pounds are annually produced : 3,000,000 acres of land are devoted to its cultivation. About 40,000,000 pounds are annually imported to the United States, and about 100,000,000 to Great Britain. It is certainly used among 600,000,-000 of people, and the commerce in it has been the source of enormous fortunes.

Green tea differs from black chiefly in its method of preparation. The popular belief that it contains poison, that it is roasted on copper, is not entirely true, though it may be that drugs are sometimes used to give tea a green color.

Paraguay Tea, or Maté, is made of the dried leaves of the Brazilian holly, and has been used in South America no one knows how long. It much resembles China tea, but is much more powerful in its effects. It is used among 10,000,000 of the South Americans.

Abyssinian Tea, or Chaat, is another poor substitute for the tea of China. It resembles China tea in its power to keep one awake, but is much inferior. The leaves when fresh are very intoxicating.

There are between thirty and forty other substitutes for China tea which have been and continue to be variously used.

COFFEE.—The coffee berry is the product of a tree that originally grew wild among the rocks of Southern Abyssinia, and in that country a beverage has been prepared from it from immemorial ages. It also grows wild in Siberia.

From Africa it was introduced into Persia about 875, A. D., and found its way into Arabia in the latter part of the 15th century.

Its **first** use in Arabia was not as a **beverage, but** merely to keep students awake at night.

About the middle of the 16th century **we hear of it in Constantinople, and along the shores of the Medi-** terranean.

Ramwolf, a German, speaks **of it in 1573** ; and Burton, in his Anatomy of **Melancholy, in 1621,** describes **it** in these **words:** "The **Turks** have a drink called **coffee, (for they use** no **wine,)** so named **of a berry as black as soot,** and as bitter, which they sip **up as warm as they can** suffer, because **they** find by experience that **that kind of drink, so used, helpeth digestion and pro-** **moteth alacrity."**

In 1652 a Greek servant of a Turkey merchant opened the first coffee-house **in London.** About the **same time it was** introduced **into** France, against **the severe opposition of** Madame Sévigné, whose influence at that **time was very** great ; and **in London also it met** with many unkind words. **In Egypt and in Turkey government and religion were** arrayed against **it.** The prophet Mohammed had forbidden **coal : coffee** was **coal, and was** therefore forbidden. **This strange logic could not stand,** and was abandoned.

So rapidly did the use of **coffee extend in** Europe that **eight years after its introduction it became a source of** revenue.

Cocoa (*chocolate*).— Cocoa is the seed of **a** very beautiful tree that **grows in the** West Indies **and Central America.** A beverage **prepared from this seed,** under **the** name of "chocollatl," **has been used** in Mexico **from** unrecorded time. **It was found by** the Spaniards **in Mexico, and** was thence introduced **into** Europe about 1520. Linnæus, the famous botanist, was so charmed with it that he gave it the generic name theobroma—food of the gods. It is used, probably, among 200,000,000 of people.

CHICORY.—Chicory is the root of a very familiar native weed. It is considerably used as a substitute for coffee, and very frequently indeed is one of its adulterations. It is used sometimes by the poor, who cannot afford the luxury of coffee. The very rough estimate has been made, that it is used among 40,000,000 of people.

TOBACCO.—Tobacco, a plant that is fully familiar, it is said was known of in Asia many centuries ago, but was not smoked or chewed. It was discovered with America, was introduced into Portugal by the ambassador of France, Nicot, (and hence nicotine,) and in England was popularized by Sir Walter Raleigh.

It was introduced into Turkey and Arabia in the early part of the 17th century.

The Chinese, of course, had tobacco before anybody else.

At the present time, 4,000,000,000 pounds of tobacco are raised annually, nearly four pounds for every dweller on the earth.

The aborigines of America were accustomed to smoke through a hollow cane, and snuff-taking was centuries ago a habit of the Peruvians.

The Aztecs of Mexico even smoked cigars, and pipes have been found in the old Indian mounds.

Tobacco, like alcohol, coffee, etc., has had to fight many hard battles.

Pope Urban VIII. thundered a bull against smoking

and snuff-taking, and King James I. of England, wrote an ugly counterblast against it. Amurath IV., Sultan of Turkey, made the use of tobacco a capital offence, and another Sultan ordered that every one who was caught in the act of smoking "should have his nose pierced with his pipe."

In Russia, one of the Czars punished smoking with bastinado and cutting off the nose.

Abbas, the Shah of Persia, even in the beginning of the 17th century, made proclamation that " every soldier in whose possession tobacco was found should have his nose and lips cut off, and afterwards be burnt alive.

This spirit of persecution crossed the ocean. About the middle of the 17th century the authorities of Massachusetts ordered " that no person shall take any tobacco publicly, and any one shall pay one penny for every time he is convicted of taking tobacco in any place."

OPIUM.—Opium is the dried juice of the poppy. It is chiefly raised in India, and is habitually used among about 400,000,000 of the inhabitants of the East.

The estimate has been made that there are in this country nearly 150,000 habitual opium-eaters. This estimate is of course nothing but a guess ; but it probably does not exceed, more likely falls short, of the truth, since the habit is one that can be indulged in great secrecy. All druggists know that in every community there are more or less who buy their opiates as regularly as others buy their tobacco.

The habit has been encouraged in this country both by the great frequency of neuralgia and other painful nervous diseases, and by the accepted views on total abstinence.

In Eastern Asia, opium, in the form of chewing and smoking, is as universal as the use of tobacco in Europe and America.

HEMP (*haschish*).—Hemp is a resinous exudation of a plant which grows in India, Persia, Arabia, Africa and Brazil. It is a plant of wide range, and is very freely used.

The Hindoos eat it and smoke it far up the slope of the Himalayas ; and among the Hottentots of South Africa we find it under the name of " dasha," used for the purposes of intoxication.

The Bushmen visiting London, amazed their entertainers by smoking dried hemp in short pipes made of the tusks or teeth of animals.

The follower of Mohammed, deprived by his religion of wine, finds consolation in hemp, and therefore we find it abundantly used among the Moors of Northern Africa.

It is a product of both hemispheres, and the Indians of Brazil know its value and delight in it.

It is estimated that hemp is used among 300,000,000 of people.

Haschish is prepared by boiling the leaves and flowers of hemp with water, and mingling butter, cloves, nutmegs, mace, etc.

The effect of hemp on the brain, is in some respects different from that of any other narcotic.

COCA.—Coca, which must not be confounded with the beverage cocoa or chocolate, is the leaf of a flowering plant that grows in the tropical regions of South America.

Popular as it is among the Indians of South America, it seems never to have been formally introduced into Europe.

It is estimated that 15,000,000 pounds of coca are produced annually, and that it is used among 10,000,000 of the human race, and almost exclusively in South America.

Coca, like many other narcotics, has had its battles. The Spanish conquerors denounced it, and brought all the powers of the Church against it. It was declared to be "an illusion of the devil," a " worthless substance, fitted for the misuse and superstition of the Indians." All these efforts were in vain. The planter found that the hands would do more work with coca than without it, and avarice, and passion, and habit, prevailed over the Church.

BETEL NUT.—Betel nut is the seed of a species of palm, that is cultivated in India, Malabar and Ceylon, where it is chewed like tobacco. It is to Asia what coca is to South America.

It has been estimated that 500,000,000 pounds of betel nut are annually produced, and that it is used among 100,000,000 of people.

One effect of its habitual use is to give a red color to the mouth, teeth, and lips, in which the native takes pride, but which to a European is at first infinitely disgusting.

LETTUCE.—The juice of lettuce somewhat resembles the juice of the poppy—opium. Like opium, it allays

irritability and induces sleep, as those who are fond of
lettuce salad well know by experience.

SYRIAN RYE.—The seeds of the Syrian or steppe rye is
used by the Turks as a stimulant and narcotic. They
are capable of producing intoxication, but their use is
not very extensive. They are used merely as a substitute for hemp and opium.

BULL's HOOF.—The flowers of this plant mixed in
powder are used in Jamaica. It has been nicknamed
the *Dutchman's laudanum*. It is one of the substitutes
for opium.

SIBERIAN FUNGUS, which much resembles our mush-
room, is a native of Kamtschatka. Like some of our
mushrooms, it is poisonous when taken in large quantities. It is chewed like tobacco.
Other stimulants and narcotics are *Sweet Galle*, which
the Swedes have used to give bitterness and strength
to fermented liquors, and the *Rhododendron Arboreum*
which is eaten by some of the people of India.

THIRD DIVISION OF STIMULANTS AND NARCOTICS.

The stimulants and narcotics of the third division
are less in number than those in the first and second,
and in a hygienic view less important. Their use is
chiefly medical. In very moderate doses, administered
internally, or applied locally, they act as stimulants.
They are also used to relieve the pain of parturition,
or of severe neuralgia.
Their main value lies in their narcotic effects, and are

resorted to when it is desired to produce anesthesia in painful operations.

It is possible to find those who form the habit of inhaling æther or chloroform, and become slaves to them ; but their number is very limited. The risk of tampering with these dangerous agents is so great, and the method of administering them by inhalation is so unfamiliar, that only the extremely reckless dare attempt the experiment.

Considering the absolute relief from pain and *ennui* that anesthesia affords, it is almost a wonder that the habit of clandestinely resorting to it is not more prevalent.

The conclusions at which we arrive from this panoramic view of the history and present use of stimulants and narcotics are these :

1. *Stimulants and narcotics have been used all over the world, and from the earliest recorded ages.*

Wine and tobacco, opium and hemp, have been used for we cannot tell how many thousands of years, and tea and coffee for many hundreds. Far back in the early childhood of India and of Egypt, we trace the evidence of the use of the wine, and further back still, in the mythical era, before the great branches of the primitive Aryans had gone forth on their mission of civilization, we trace the sacred rites connected with the " Divine Saoma."

The science of language, boldly penetrating the mist and darkness, that boundless cloudland that lies beyond recorded history, finds there no people by whom fermented liquor of some form, had not been discovered.

Tobacco is used by a larger number of the human

race than **any** other article, except salt, and wine **is** more universal **than** bread, with which it is so often assorted. **Few, if any** articles of food are so widely consumed as **tea and** coffee, and even whiskey is **more** universal **than the** potato.

Very probably there are **some** wild tribes in the world who have **no** form of stimulant or narcotic, just **as there are some who appear to have** no form **of** religion, **but such exceptions, if any there are, will certainly be merely** enough to establish the general rule.

2. *Their use has greatly extended and multiplied with the progress of civilization,* **and especially** *in modern times.*

In the mythical **and early historic ages, the** number of stimulants and narcotics that were known and used **was quite** limited. **During the past ten** centuries, since **the discovery of distillation, of** brewing, and the **spread** of tobacco everywhere on the discovery of America, their use has multiplied many hundred fold. Two great **events have combined to** increase the consumption of stimulants and narcotics with the advance **of the race.**

First, the discovery and invention of new varieties, or new modifications of old varieties.

Secondly, the influence **of commerce,** by which the products **of each clime** became the property of all.

The **ancient civilizations** knew only of home made varieties ; the **moderns are** content with nothing less than all of the **best that the** world produces.

The **commerce** of modern times consists very largely in the transportation of stimulants **and** narcotics, and there never was a time in the **history** of man when they were used **so** liberally **as at** present.

3. *They are used to the greatest extent and in the lar-*

gest variety among *the nations* who are now leading our *modern civilization.*

The most enlightened, the most progressive nations of our time, are Great Britain, Germany and the United States, and in these countries, stimulants and narcotics are used in the greatest abundance and widest variety. Next to these nations in order of enlightenment, and in order of indulgence in these substances are France, Russia, Sweden, Norway, Italy, and Spain. The semi-civilized nations, as Turkey, Syria, India, China, Japan, South America and Mexico, use some varieties to considerable excess, but have not as many varieties. and do not, on the whole, use as great a quantity of stimulants and narcotics as the nations who are at the head of civilization.

The purely barbarous races and tribes use at most but one or two varieties, and as a rule, to but little excess. Africa seems to use less than any other continent.

CHAPTER II.

HOW SHALL WE ASCERTAIN THE EFFECTS OF STIMULANTS AND NARCOTICS?—This question may be best answered by first indicating some of the methods by which we *cannot* ascertain the effects of these substances.

We cannot ascertain their effects by consulting exclusively our hopes and fears. The emotions, as propelling forces, are powerful and indispensable ; but when they attempt unaided to determine questions of science, they usurp the throne of reason and become the worst enemies of truth. It is proper that all the faculties should join in the pursuit of truth, but each one should keep its appointed place.

The question is pre-eminently one of fact, and, as in all other questions of fact, the last appeal for decision must be made to reason ; and we must follow truth wherever it leads, even though every step smashes some favorite idol.

We cannot find out by chemical or physiological experiment. The sciences of chemistry and physiology are yet too undeveloped to be of very important service in enabling us to answer the great fundamental questions, what shall we eat ? and what shall we drink ?

We know something of the structure of the human body; something of the chemical constitution of the leading stimulants and narcotics; something of the relation of chemical substances to nutrition; but not enough of any of these departments to build up anything like an exact or complete science of diet.

Indeed, so far are we from reaching this long sought-for goal of wisdom, that if we were forced to wait for chemists and physiologists to determine by experiment what we should eat and drink, we should starve to death before even the simplest substances could be scientifically proved to be innocuous. I suppose that if chemists and physiologists were to unite their forces, and attempt to prove by experiment alone, the part that pure water plays on the human economy, their task would be one of more than usual difficulty, and a stubborn doubter could raise more queries in one moment than they could answer in a century.

Messrs. Lallemand, Duroy and Perrin think they have proved that alcohol, administered in large doses, is eliminated from the system.

Baudot thinks that he has proved that when alcohol is given in small doses, it is not eliminated.

Messrs. Bocker, Lehmann and others think they have shown by experiment that tea, coffee, alcohol, etc., retard the metamorphosis of tissue.

It may be that all these observers are partly right; it may be that they are all partly wrong.

I honor the patience and carefulness and scientific enthusiasm of these observers, but I see nothing in their conclusions except that we as yet know almost nothing of physiological chemistry. Even if they all agreed instead of widely disagreeing, and their evi-

dence were reinforced by observations without limit,
still the skeptical thinkers would hesitate long before
allowing the solution of the great question of stimulants and narcotics in all its vast relations to hygienic
morality, political and social economy, and science in
general, to rest on such experiments.

PHYSIOLOGICAL CHEMISTRY IN ITS INFANCY.

There is scarcely a theory in physiological chemistry
that may not be flanked to-morrow by some other theory, which in turn may share the same fate. Experi-
menters swallow their own experiments. Most of the
brain force of scientists in these departments is spent
in disproving what others have proved, or perhaps in
eating their own words. This, then, is not the best
road by which we shall arrive at the truth. Apparently the shortest and surest way, opening well, with
broad and firm pavements, and alluring borders, it
soon becomes thick with pitfalls, into which at every
step one is liable to stumble, and finally runs into an
interminable jungle where the thick foliage shuts out
the light, and not even a footpath can be found as a
guide to the traveller.

Physiological chemistry, youthful as it is, is not to be
despised; rather is it to be highly honored, but not by
assigning to it tasks that it cannot fulfill. After we
have once found what substances are best to eat and
drink, experiments may explain somewhat their action
and their relation to the human economy, and it may
enable us to use these substances more intelligently.

In this capacity, as an explainer or harmonizer, it
serves a good purpose; but even here it blunders at

every step ; but blunders here are of less moment, since they can do much less practical evil.

With all the sciences progressing every year ; with humanity increasing in capacity and in force ; with opportunities multiplying for the exercise of this capacity and force, he would be a bold man who should predict that ages hence physiological chemistry might not approach to completeness ; but he is rash and wild who declares that it has already attained that eminence.

Again, we cannot learn the effects of stimulants and narcotics by observing their effects on the lower animals. In many features the lower and higher animals strikingly concur ; but such concurrence is not uniform.

All species of animals have their peculiar sustenance. What is food for one is injurious to another. A man would starve on the grass and clover that keep his oxen well, hale and strong.

Every reader of Dr. Livingstone remembers the famous tsetse fly that was so fatal to the ox, the horse, and the dog, but was harmless to man and wild animals.

Absurd beyond all degrees of absurdity, therefore, is the habit of inquiring of dogs and cats and swine, what we shall eat and drink. With even more reason might we inquire of them wherewithal we should be clothed.

And yet, during all this contest on the subject of stimulants, dogs, and swine, horses, and cows have been summoned to the witness stand and required to give their opinion concerning the effects of these substances.

These animals have pretty unanimously and very emphatically expressed themselves in opposition to many of the stimulating and narcotizing substances that are so popular among their human brethren. They have testified that whatever their tastes might be in the future, after the theory of natural selection had had larger sway, they now have no desire whatever for rum or tobacco ; that they despise the taste of the one and the odor of the other. Even the hog, the most greedy of animals, prefers a swill-pail to a demijohn, and would rather roll in the mud than take a whiff of the finest Havana ever made.

These facts are undeniable, and they have been quoted with warm approval as settling the whole great question of the use of stimulants and narcotics among the human race, and that, too, by men who would feel that their arrest would be justifiable if they should walk the streets with no other protection than nature afforded them ; and would complain bitterly if they were obliged to subsist on dry husks, or hard corn, or even on swill milk and green grass.

EXPERIENCE THE ONLY GUIDE.

The one and only way by which we can learn the effects of stimulants and narcotics on the human system is by *experience ;* by trying them on a large number of individuals, and observing their effects over a long period of time.

This is experimenting, but it is experimenting on an enormous scale, and by the only true method. It is a method not unattended with difficulty, for it requires the accumulated observations of many individuals at different times, in various climates, and with all sorts of environments, and their observations must be compared, in some cases contrasted, and the unknown and unknowable quantities must be eliminated, and out of this tangle an approximately correct solution is now obtainable, for we have at command something of the accumulated experience of the world, most of which, during the past two or three centuries is quite available.

Let us approach this subject calmly and leisurely, that we may make no false steps, and survey thoroughly the whole field.

Let us approach it reverently, for we are seeking to enter the temple of truth.

We can best arrive at the whole truth on this subject by studying all forms of stimulants and narcotics collectively. What effects are common to all, and how are these effects, in the case of each prominent variety, modified by various conditions?

First of all we recognize the fact that all stimulants and narcotics, so far as they have been studied, contain active principles that are more or less poisonous, and it is mainly for the sake of these active principles that they are so eagerly sought for.

In making the statement that all these substances contain poison, it is necessary to present a definition of the word poison, and accordingly I adopt the one given by Webster—"any agent capable of producing a morbid, noxious, or dangerous effect upon anything endowed with life."

The qualification "in comparatively small quantities, or in quantities not very bulky," might be added, but it is not essential to the correctness of the definition.

By this definition all stimulating and narcotizing substances contain poison, for they all contain active principles which are capable of producing not only noxious, but actually dangerous effects on the human system.

Whether these poisonous active principles exert the poisonous effects of which they are capable when administered to anything endowed with life, depends on a variety of conditions.

It depends :

1. On the quantity given. A small quantity may do good, or at least cause no harm, while double the amount injures or kills. This is true of arsenic, phosphorus, strychnine, prussic acid, veratrum viride, aconite, opium, and indeed of nearly all the best known and most virulent poisons. On the other hand, calomel, in small doses, may be poisonous, and show its poisonous effects by causing salivation, while in larger doses it works great benefit. This distinction has long been recognized by the profession, and is not unknown to the masses. Calomel is in bad odor at the present time, and the force of this illustration will not be as fully realized as it would have been fifty years ago.

2. On the substances with which it is combined. The virulence of poisons may be much increased or diminished by combining them with various substances. Opium is less harmful if given with belladonna ; and hydrate of chloral, it is said, can be given with less caution if at the same time strychnine be used.

Phosphorus, in the form of phosphates, and as combined in fish and flesh, can be used more freely than when not so combined. Theine, and caffeine, and alcohol, and nicotine are all mollified more or less by the substances with which they are respectively associated in tea, coffee, wine, and tobacco.

3. On the individual temperament. This is simply a

statement of the truism that what is one man's food, is another man's poison, and will be conceded without dispute.

The leading active principle of fermented and distilled liquors is *alcohol*; of opium, *morphine*; of hops, *lupuline*; of tobacco, *nicotine or nicotianin*; of tea, *theine*; of coffee, *caffeine*; of cocoa, *theobromine*; and all are poisonous. And it may be inferred that the less known forms that have not been chemically examined, do yet contain poisons more or less analogous, since their effects in the main are identical.

Different poisons, for inexplicable reasons, have a preference for different parts of the body. Arsenic, for example, affects chiefly the mucous membrane of the stomach and bowels. The poisonous active principles of stimulants and narcotics so far as is known, affect chiefly the entire nervous system, the brain, spinal cord, and great sympathetic ; either attacking all at once or in succession, or spending their strength mostly on one.

From the weaker forms, as theine and caffeine, to the stronger, as nicotine and morphine, through all the grades, they are capable of producing paralysis of the great nervous centers.

Dr. R. Amory, of Boston, who has made many experiments with theine and caffeine, found that animals even of considerable size could be paralyzed and killed by these drugs, in a comparatively short time, and in doses not enormously large. Twenty grains of theine introduced into the crop of a pigeon, killed it in less than two hours. He gave ten grains of theine, on a piece of meat, to a fasting dog. In three quarters of an hour the dog walked the room, " panting and holding out his tongue, respiration being much hurried." These attacks continued by intervals of about fifteen minutes, for two or three hours. On the following day the animal was perfectly well.*

Thirty grains of caffeine killed a terrier-dog in about twelve hours or less.

Twenty grains of caffeine introduced into the stomach of a rabbit, through the walls of the abdomen, killed it in twelve hours.

Ninety grains of theine introduced into the stomach of a dog in the same way, killed it in about an hour.

The conclusion of Dr. Amory was that these drugs, which are very similar in character, chiefly affect the spinal cord, as shown by the convulsions that precede death.

The discovery of the fact that all these substances contain poison, sheds no light on the query whether they are or are not injurious to the system, when habitually used ; for, as has been stated, the question whether their poisonous effects are developed, depends on a variety of conditions.

Phosphorus is one of the most virulent of poisons, but it is found in fish and meat ; and partly for this reason is it that fish and meat are good diet for brainworkers.

There is poison in the garden lettuce and in the luscious vegetable. There is poison in the air we breathe ; in the water we drink ; and in the milk which the infant draws from its mother's breast.

Poison, therefore, comes to us in the stormy wind and bursts out of the earth with the fruits and herbs and

---

* Dr. R. Amory, Boston Med. and Surg. Journal, p. 266, May 28, 1868.

flowers. Go to the table of the *bon-vivant*, or to the hard board of the peasant, and we cannot escape it. It lurks in the loaf of bread, as well as in the " red, red wine ;" without it we could not be born, for we are made of it ; without it could not exist for one day or hour.

Of the fourteen elementary substances that enter into the structure of the human body, nearly all are poisons, and the waste of all these must be supplied by our food and drink.

Common salt is a poison, and in doses not very bulky, and in more instances than one it has been known to destroy life ; but it is found in many of our articles of food and is the most universal nutriment of man. The most zealous hater and fearer of poison cannot partake of an ordinary meal, nor enjoy a bath at home or in the sea, nor drink a glass of water from the spring, nor draw a single breath without taking into his system poisons far more potent than those he hates and fears.

Deprive drinking water of its poison, and you make it insipid ; if a meal were prepared from which all poison should be extracted, it would be tasteless and unnutritious, and the chief advantage of salt over fresh water bathing lies in its poisonous ingredients, and the tonic effects of the sea air are largely due to the poison with which it is laden.

To say that any substance is in general a poison, gives, then, no clue to the question whether it does or does not serve a purpose in the animal economy. Some can be used habitually with profit, and pleasure ; others cannot.

Whether a substance abstractly poisonous exerts an injurious or a beneficial influence in any given case, depends as we have seen on its dose, its combination, the temperament of the individual, on the age, sex, and state of health.

All these questions experience, and only experience, can decide.

Although stimulants and narcotics are valued chiefly for the poisonous active principles which they contain, many of them contain also other principles that are more or less positively nutritious, or substances that in some way affect the system.

Tea, for example, contains a volatile oil, starch, gluten, fat, tannic acid, woody fibre. *Coffee* contains the same substances, with less gum and sugar, less gluten, less tannin, and more woody fibre. The gluten in tea and coffee is positively nutritious, and if tea leaves and coffee berries were eaten in mass like peas and beans, which are valued for the gluten they contain, they would afford substantial nourishment and sustain life, for in tea leaves the percentage of gluten is about as great as in beans. There are about ten grains of theine in one ounce of tea.

The milk and the sugar that are used with tea and coffee contain also positive nutriment ; but after all, it is mainly for the effects of theine and caffeine that these beverages are so freely used.

Wines contain, besides alcohol, grape sugar, in large or small quantity, a free acid, cream of tartar, tartrate of lime and œnanthic ether, to which they owe their peculiar and distinctive flavor or odor.

It will be seen, therefore, that wine is a very different substance from a simple mixture of alcohol and water.

Ardent spirits are sought for almost solely on account of the alcohol which they contain ; but wines are desired for their flavor, for their sweetness or acidity, for the positive as well as negative nourishment which they afford.*

Chemical analysis shows that the juice of the grape contains water, sugar, acidulated tartrate of potash, tartrate of lime, phosphate of magnesia, muriate of soda, sulphate of potash; sometimes also gallic, citric and malic acids.

Opium contains a certain small proportion of meconic acid, fat, resin, gum, mucilage, and about twenty-five per cent of gummy extraction.

Beer contains a small quantity of solid matter called malt extract, consisting of sugar, gluten, and bitter substances of the hop and of mineral matter, averaging from four to eight per cent. The nutritive matter of milk is about twelve per cent; it seems that beer is at least one half as nutritive as milk. It is, therefore, positive as well as negative food.

---

* The following table, showing the percentage of alcohol in the different liquids, I take from Cameron's work on *Food and Diet.*

Percentage of Alcohol.		Percentage of Alcohol.	
Rum	60 to 75	Chablis and Sauterne	8 to 12
Whiskey	54 " 60	Rhine wines	7 " 15
Brandy (British)	50 " 60	Champagne	7 " 13
French brandy	50 " 55	Burgundy	8 " 12
Gin	48 " 58	Moselle	8 " 13
Port wine	14 " 24	Ale	6 " 9
Sherry wine	14 " 27	Cider	5 " 9
Roussillon	11 " 16	Porter	4 " 7
Claret	9 " 14	Beer	2 " 4
Hungarian wines	9 " 15		

Ardent spirits, besides alcohol, contain a very little sugar, small quantities of oily substances, and certain æthers that give them their distinctive flavor. Rum, for example, contains butyric æther; gin contains oil of juniper, by virtue of which it affects the kidneys; and whiskey contains more or less fusel oil.

Cocoa contains a larger per centage **of starch and** gluten than either **tea** or coffee, **and more than** fifty per cent **of fat.** The *theobromine* **of** cocoa closely **resem-** bles theine **and** caffeine in its nature and its effects.

Cocoa is much more nutritious in a **positive way than tea** or coffee, and may indeed take rank with ordinary articles **of food.**

They **all** *have* **the** *power* **of** *sustaining the system,* **and** *within certain limits, can take* **the** *place of ordinary food.*

I have called these stimulating and narcotizing substances NEGATIVE FOOD, in **distinction** from our ordinary diet or *positive* food.

I call them *negative* food because it **seems to be** tolerably well established by the **experience of** mankind that they **sustain the system in a** *negative* way ; not so much by supplying **the waste, as by making it less, and** enabling the system **to feed on itself, and** aiding us in economizing **our** vital resources. Slaves **to alcohol,** opium, tobacco, or even **tea** and coffee, are **usually mod-** erate eaters, **for** the reason that the **immense quantity of** negative **food** which **they** take, habitually, in **their** favorite stimulant or narcotic, makes **less imperative the** need for positive nutriment.

In the light **of this fact we see why** it is that civilized man makes more liberal use of **s**timulants and narcotics, than the uncivilized. In civilization the expenditure of force is vastly **greater** than in barbarism, because **the brain** especially is more active ; **to compensate for this** expenditure, to **retard** the waste **of** tissue, **or** at least to sustain the body amid the cares, toils, and pressures incident to advanced civilization, **men** resort not only **to a** more liberal and abundant variety of food than savages use, but also *most* employ a *wider range* of stimulants and narcotics, even though they may not use any one to so great excess as do certain wild or semi-enlightened nations. Hence result deplorable consequences.

The poor, and ignorant, and idle classes that have almost always been found among highly civilized people, are brought into the presence of the same variety of stimulants and narcotics that the civilized brain-working orders devise and employ ; and in want of sufficient moral force and elevation to use these substances with decorum, they fall into the habit of indulging in them to enormous excess.

SUSTAINING POWER OF TEA AND COFFEE.

I have always observed that tea and coffee drinkers eat less than those who wholly abstain from these articles.

I first made this observation years ago, in my boyhood, when I had, to say the least, no theories to prove on the subject, and when I approached the investigation untrammeled by prejudice.

Those who alternate between total abstinence from all these things, and moderate use of them, find very often that, when they totally abstain, they eat more of ordinary food.

Thousands there are to whom a cup of coffee is the chief feature of their morning meal.

Soldiers and Sailors regard the deprivation of coffee as a greater affliction than short rations of meat or bread.

The Gallæ, a tribe of Africa, in their long wander-

ings, sometimes take nothing with them except a bag
of coffee and butter, done up in bags about the size of
a billiard ball, and one bag-full suffices them for a day.

SUSTAINING POWER OF TOBACCO.

The power of tobacco to sustain the system, to keep
up nutrition, to maintain, and to increase the weight ;
to brace against severe exertion, and to replace ordinary
food, is a matter of daily and hourly demonstration.

I have known convalescents from fevers to be sup-
ported for long periods very largely by their quid or
pipe.   Sailors and soldiers in the war, relished the
tobacco rations even more highly than their coffee.
There are thousands who, in crises requiring great
exertion, forego their meals and increase their pipes or
their chews.

SUSTAINING POWER OF ALCOHOLIC LIQUORS.

The sustaining power of alcohol in health or in disease is very great.   A lady friend of mine, on account
of the very severe illness of her son, was obliged to
confine herself very closely to the duty of nursing, with
great loss of meals and of rest.   Owing to intense and
long continued anxiety, she became so reduced that
she could eat scarcely any solid food whatever, and for
more than two weeks she lived almost exclusively upon
lager bier, of which she took some half dozen bottles
daily.   This, with now and then a bit of cracker, was
her only nutriment.

At the end of that time she was not very greatly reduced in weight, and retained a fair share of vigor.

Dr. Anstie who has written so ably on this subject,

cites a number of remarkable illustrations of the **power** of alcohol to sustain life.

An old man of eighty-three years, affected **with** chronic bronchitis, lived for about twenty **years on a** bottle of gin a day, and a small finger-length of bread a day. He used no tea and no coffee, but smoked his pipe.

A young man, with disease of the heart, lived two years almost entirely on **brandy**, since nothing else would remain on his stomach. At first, six ounces ; subsequently, a pint was given him. " He kept his flesh and good spirits nearly to the last."

One child lived on Scotch ale for a fortnight, and another, suffering from marasmus, lived for three months on whiskey and water alone.

One lady about twenty-five years of age, lived "solely on bitter ale and brandy and water " for twelve months, at the end of which time she was greatly prostrated, but was not emaciated.

Such exclusive dependence on alcoholic liquors is of course to be deprecated, except in peculiar emergencies where nothing else can be obtained, or nothing else can be tolerated. Used in this way, even the milder forms of stimulants will more than likely bring on or aggravate serious organic diseases.

There are crises in life when this power of alcohol to sustain the system may prove a signal blessing

SUSTAINING POWER OF COCA.

The power of coca to sustain the system during abstinence from ordinary food and amid severe labor, is something remarkable. A large number of trustworthy travellers agree that not only the South American Indians, **but** foreigners residing in South America, can **perform an** immense amount **of** labor with **no** other sustenance than that derived from coca, **and they** further agree that this is not followed by any unpleasant reaction.

Von Tschudi states that an Indian sixty-two years **old worked for** him five consecutive days and nights with very little sleep, with no ordinary food, and **no** sustenance except that **derived from** the coca that he chewed.

At the end **of the** time he was **well** and able to undertake a long journey.

Mr. Whittingham, a surgeon **of Callao in 1850, says that two men** were buried in a mine eleven days before they could be dug out, and kept themselves **alive by** coca and the raw hide of their sandals.

SUSTAINING POWER OF OTHER STIMULANTS AND NARCOTICS.

In the Philippine Islands, travellers and day laborers depend on their rolls of **betel,** as the South American Indian depends on his coca. Many would give up their meat **sooner than this.**

The couriers of Tartary and India will travel **for** days on nothing but a small piece of opium, with a little rice and dates. So highly is it prized as a sustainer **of the** system amid fatigue, that in Mohammedan countries the cakes that are made for **the** use of travellers, are stamped with the Turkish legend : "Mash Allah,"—"the gift of God."

The Cutchee horseman even shares his opium with

his horse, and both bear their fatigue with amazing ease so long as their stock holds out.

Their effects are modified by climate, the season of the year, and the daily and hourly changes in the atmosphere.

In regard to the relation between climate and the effects of stimulants and narcotics, these two propositions can, I think, be pretty clearly established; *first,* that they can be used with most benefit and least injury in the countries where they are naturally produced; and *secondly,* that none of them can be used as freely in countries where the air is *dry* as in those where it is *moist.*

Tea is a product of a temperate clime, and can be, and is liberally employed as a beverage in temperate regions in both hemispheres. Although used more or less in all civilized and semi-civilized lands, it is yet most freely used in cold and temperate latitudes. Next to the Chinese, the Russians are the greatest tea-drinkers of the world; after the Russians come the Americans and the English. In France, Italy, Turkey, Spain, South America and the Southern United States, tea is to a considerable extent replaced by coffee. In making these distinctions, mankind but follow their instincts and experience. Coffee is a natural product of warm climates, and is better adapted to the inhabitants of those regions than tea; but in cold and temperate countries, coffee, for some reason which we do not understand, is very frequently injurious.

In the northern United States especially, there is an increasingly large number in both sexes, and of all ages, who cannot habitually use coffee, and I daily meet with patients who have found out by experience, without the aid of any physician, that they must forego the enjoyment of this most delicious beverage.

FREE USE OF COFFEE IN THE SOUTHERN STATES.

In the Southern States it is the custom with very many to drink coffee several times a day, and not merely at the morning meal. In the Northern States, this habit of taking coffee at all the meals, at lunches, and in the intervals of meals, is very rare. There is no reason for believing that the average result of the free coffee drinking in the Southern States, is injurious either to the digestion or the nervous system ; but if the custom were transferred to the Northern States, and persisted in, we should become a nation of the worst of invalids.

Individuals who, in the latitude of Charleston or New Orleans, can drink strong coffee several times daily, must here tremblingly confine themselves to a single cup, or very likely deny themselves entirely.

While residing South, for a year and a half, during the war, I gradually fell into the custom of drinking strong coffee several times daily. Not having been accustomed to using even one cup a day in the North, on account of the great disturbance that it caused, I, at first, feared unpleasant symptoms from such apparent excess, and was surprised to find that I was benefited by the indulgence ; that it enabled me to bear with less difficulty the depressing heat of the climate.

On my return North I found that I was obliged to return to my old habit of total abstinence, or suffer from nameless horrors.

I have met with other instances equally striking.

The liberal dispensation of coffee was one of the best investments that was made by the department during our war.   It compensated not a little for the necessary deficiences and monotony of the solid rations.   It prevented, if it did not cure, malarial and other diseases.

In many other warm countries, as Spain, South America, Italy, and Turkey, coffee is used with a freedom which in colder climates would not be tolerated.

The air of England and of the continent of Europe is far moister than ours, and for that reason nearly all forms of stimulants and narcotics can be used more liberally there than here.*

It is the experience of the majority of smokers and drinkers that free indulgence in the latter part of the day—in the afternoon and evening—is better borne than in the morning. Especially is this true of tobacco, which among moderate smokers is usually reserved for the evening, and the hour after the last meal.

I have been told by habitual smokers and drinkers that they can indulge in their favorite stimulants more largely in damp weather than in dry ; and many who are exceedingly susceptible to the effects of these agents, will bear witness that certain times of the day are more favorable for liberal indulgence than others ; that the effects which they experience are strangely modified by the seasons of the year, the hours of the day and the state of the atmosphere. The fact that there are some who appreciate these differences, suggest the probability that very many others are similarly influenced, though not to a sufficient extent to excite their attention.

---

* This subject will be discussed more in detail in Chapter III.

The natural wines of any country are generally beneficial to the inhabitants of that country; but the
strong distilled liquors can hardly be regarded as a natural product, and in large, and very frequently in
small quantities, are injurious in all climates.

It is quite generally held that gouty patients are
injured by sour wines, and this is probably the case in
England and America; but on the banks of the
Rhine, where the acid wines are made, and where they
are drunk by everybody as freely, almost, as tea or coffee with us, gout is by no means a prevalent disease.

Ardent spirits seem to be more likely to injure in extremely hot or extremely cold than in temperate climates. It is quite certain, however, that in moderately
warm weather, such as we have in summer, when there
is abundance of sensible perspiration, or in the latitude
of our southern States, more alcohol can be taken without causing intoxication than in cold seasons. It is
well known, also, that the different varieties of ordinary
food, as meat, vegetables, fruit, etc., vary in their
effects in different climates, and at different seasons of
the year.

The diversity in the effects of stimulants or narcotics
at different hours of the day, may perhaps be explained
by the facts concerning the variations of atmospheric
electricity, of which I shall speak in Chapter III.

*Their effects vary with the race and the progress of civil-
ization.*

The cataleptic state produced by hemp seems to be
peculiar to the Eastern races; it certainly is not observed by those of the Europeans or Americans who
experiment with the drug.

It is hard to discriminate the influences of race and of climate in causing tolerance or lack of tolerance of stimulants and narcotics, since the different races are very apt to live in different climates. Judging from comparative observations in this country, it would seem that the Germans bear more of these substances than any other race—more even than the English and Scotch. It must be allowed, however, that the Germans use less of ardent spirits, and more of mild beers than most other civilized people. In tolerance of tobacco they are greater than any other civilized people.

It follows from these considerations that their effects must be modified by the progress of civilization.

That the type of constitution undergoes important changes in the progress of mankind from barbarism, through the various stages of refinement, is now well established, both on *a priori*, and on historical grounds, it follows therefore, from what has been previously stated, that civilized man must be very differently susceptible to stimulants and narcotics, from the savage ; and that among the grades of advancement, the laborious brain-worker will be affected by them very differently from the one who subsists partly or entirely by his muscles.

We know that our fathers could bear more alcohol than their nervous, though longer-lived descendants ; and to my own mind it is sufficiently clear that, among the brain-working class of our population, the proportion of those who can with impunity make large use of tea, coffee and tobacco, is yearly diminishing. We must be more wary, more solicitous of ourselves than our fathers were ; and though descended from them, and under pretty nearly the same climate, we are not like them. The change has not been altogether for the worse, but in some respects for the better. If they had fewer nervous diseases than we have, they suffered more from inflammations, and from rapidly spreading and rapidly fatal epidemics.

The diet on which our fathers grew hale and strong, lies heavily on our stomachs, torturing us with indigestion and frightful neuralgia ; just so the whole range of stimulants and narcotics must be used by us, very differently from what they were by them, else we suffer.

Their effects **vary** *with the individual temperament.*

Nothing in science is better established **than** that individuals **vary** widely in their susceptibility to the different articles of nourishment of all kinds.

Tobacco has widely opposite effects with different individuals. Some it fattens, others it withers; for some it causes dyspepsia and constipation; for others it relieves dyspeptic symptoms, and relaxes the bowels ; **for some** it produces sleep, for others, wakefulness ; some temperaments it arouses to intellectual brilliancy, others it muddles and stupefies ; **on many** its effects are to calm, **to** soothe, and to produce a sweet and mild oblivion ; **on** others it brings all the horrors of extreme and **painful** nervousness. Some like a smoke before going to battle, **to** brace their strength and courage to unusual effort ; others like it after the battle, to calm their nerves and soothe them to slumber.

The effects **of** a sudden breaking off **of** the habit of using tobacco, vary remarkably **in** different individuals. Some, at once, improve in health, increase in flesh and strength ; others deteriorate, becoming thin and **weak.**

Persons in whom the nervous diathesis exists, especially vary in their susceptibility, not only to tobacco but to all other forms of stimulants and narcotics, fermented and distilled liquors, tea, coffee, opium, etc.

The usual tendency of coffee is to produce wakefulness ; and those who are obliged to sit up late at night, often find that a cup of strong coffee enables them to keep awake ; but there are those whom the same beverage disposes to slumber, and who therefore are wont to indulge the luxury just before retiring. The same directly opposite effects in regard to sleep, come from tea and alcohol, and even opium itself, which has always been regarded as one of the surest of sleep-producers, frequently causes extreme wakefulness. One of the most uniform effects of opium is to produce constipation ; but it has been known to cause the opposite condition to which the term " opium diarrhœa " has been applied, and my attention has been called to a striking instance of this that occurred in the practice of my friend, Dr. Conkling of Brooklyn.

The different forms of alcoholic liquors, wines, cider, beers and ardent spirit are so capricious and uncertain in their effects on different individuals, that physicians when they prescribe them for the first time to a patient with chronic disease, usually feel that they are trying an experiment. Sleep and wakefulness, headache and relief of headache, flushing of the face and a feeling of heat in the extremities, indigestion and relief and prevention of dyspeptic symptoms, constipation and looseness, fat and emaciation, strength and weakness, stupidity and brilliancy—all this vast variety of opposite symptoms may and do result in different constitutions from the use of alcoholic liquors, and that, too, from the weakest as well as from the strongest forms, whether used in large or small doses. Some of the preparations of alcohol are more likely to produce unpleasant and injurious effects than others. Thus the habitual use of rum, whiskey, gin, brandy—which contain about fifty or sixty per cent of alcohol—as they are pared and adulterated at the present day and in our country, in the long run bring on more disease than they prevent or relieve, and in every way work more injury than benefit.

Tea is a beverage that can be used without injury by a larger number of people than any other form of stimulant and narcotic ; but I have known individuals whom even a single cup would keep awake all night, or produce most distressing nervousness.

Cider is one of the mildest of fermented liquors, but there are many ladies for whom half a glass is sufficient to cause headache, sleeplessness and similar nervous sensations.

Individuals are also differently susceptible to articles of diet. I have heard of a lady who was sadly dyspeptic, and who could digest nothing but sausages, and drink nothing with comfort but hard cider.

A clerical friend of mine, who has a very good constitution every way, and who is now about eighty years old, cannot eat even one strawberry without being thrown into convulsions.

I know a young man who, if he enters the parlor of a house while they are cooking buckwheat cakes in the kitchen, and all the doors are closed, will be thrown into a violent attack of asthma, and if he dares to eat as much as will cover a three cent piece, a rash comes out all over him, and yet strawberries and buck-wheat are healthful food.

SUSCEPTIBILITY OF INDIVIDUALS TO MEDICINES AND AND ODORS.

Equally strange is the susceptibility of individuals to many of our most potent and most useful medicines, as strychnine, hydrate of chloral, opium, chloroform, and æther. Hence arises the necessity of very great caution in the use of the agents ; for although but one in many thousands manifests this peculiar and dangerous susceptibility to them, yet that one may be the first patient to whom they are administered.

The odor of vanilla, so justly prized as a flavor, is said to intoxicate the laborer who gathers it ; and some, it is claimed, are poisoned by the perfumes of the pink, the rose, and other flowers.

Their effects vary with the sex. Men can bear them better than women.

This rule applies, I believe, to nearly every form of stimulant ; although individual exceptions to the law, both ways, are quite frequent.

The general law is that the more nervous the organization, the greater the susceptibility to stimulants and narcotics. Woman is more nervous, has a finer organization than man, is accordingly more susceptible to most of the stimulants.

In civilized lands, where her nervousness is most apparent, woman abstains almost entirely from tobacco, having pretty generally abandoned both her snuff-taking and her pipe. Even in countries where the vine grows, she uses a less amount of wine than man, though she drinks it, perhaps, as habitually and as frequently.

Tea is woman's great beverage, since by its mildness, and by its peculiar nature, it is the best of all adapted for her fine organization. It is used to excess by women more than any other of the stimulants, and much more than by men. It is for woman what tobacco is for man. It is her solace and her strength. It makes ordinary food more palatable, and aids in its digestion, and frequently supplies its place.

When the appetite is repelled through grief or disease ; when the organs of assimilation have lost their vigor through long abstinence, or repeated irregularities attendant on the cares of rearing a family, tea gratefully sustains the failing energies, and that too with the slightest possible tax on the organs of digestion.

Their effects vary with the age. The old bear them *better than the young.*

Those who carefully study their own constitutions, find that varieties and preparations of food which at one time of life are beneficial, at others prove injurious, and all the world knows that the diet of infants and children must be radically different from that of adults, and that adults in turn must modify their habits of eating in extreme old age. Negative food—stimulants and narcotics—is much more variable in its effects at different periods of life than positive food, and hence there is need of greater caution in using it.

IN INFANCY AND CHILDHOOD, LITTLE NEEDED.

Infants and children do not need stimulants and
narcotics, and should not ordinarily be allowed even the
weaker varieties. The young of both sexes, whose growth
is not completed, are more liable to be harmed than benefited by a free indulgence in these substances, for
these three important reasons : *first*, while the system
is growing it needs abundance of *positive* nutriment to
supply the rapid changes of tissue, rather than negative, which lessens the appetite for other food ; *secondly*, their brains are much less actively employed
than those of maturer age, and therefore have less
need of the sustaining influence of stimulants and
narcotics ; and *thirdly*, there is great danger that
they may acquire habits of over indulgence and become slaves to appetite, since at this time of life they
have neither the full development of moral character,
nor yet the manifold restraining, diverting and counteracting influences and duties that enable adults
to use them in a wise and decorous moderation.

The tendency with many parents, in this country certainly, is to allow their young children to form habits of
using tea and coffee much too early, and the tendency with young men, in every walk of life, is to acquire the habit of smoking, chewing, and drinking
strong liquors at a time when they have little need
of these substances, and when they have not the
moral force to resist their seductive and bewitching
influences.

I have long thought and have frequently stated,
that if our young people would avoid the form-

ation of habits of free indulgence in stimulants and narcotics until the growth is completed—a period which variously ranges between the age of twenty and thirty—intemperance and most of the physical disease that results from these substances, would in the course of the next generation be well nigh unknown, for habitual intemperance, like most other crimes, is usually the result of habits formed in youth.

ESPECIALLY BENEFICIAL IN OLD AGE.

On the other hand, some of these articles of which we have been speaking, which in youth are of such doubtful advantage, in the decline of life are as beneficial as they are grateful.

By gently stimulating the jaded digestion, by giving tone to the exhausted brain, by equalizing the languid and unbalanced circulation, and by economizing the wasted tissues, they beautifully and efficiently sustain the system when the desire for positive nutriment has long been blunted, and the forces of assimilation have well nigh lost their mysterious power. Thus they serve a most beneficent purpose to sweeten and prolong the evening twilight of existence, to make less perceptible the slow darkening of the lights in the windows, more gently and easy the sure descent into the depths of the dark unknown.

Surprise is often expressed by individuals that as they advance in life, they can use with benefit, or at least without injury, substances which in their previous history they could not bear.

Our bodies change continually, without regard to age. They are changed by the weather; by our occupation and duties ; by our joys and woes ; by our diet and sleep, or the want of it ; by the varied conditions of disease ; by all that makes up our life; they are changed even when we are the least subject to external influences. The elaborate chemistry of the body never rests. I am not the same individual to-day that I was yesterday ; to-morrow I shall be different from what I am to-day ; a month hence I shall be a new being ; thus, in the course of life, I personate number-less different, and even opposite physical characters. A stimulant or narcotic, therefore, which I take to-day, affects me very differently from what it did last year, because I am not the same person.

Prof. Huxley tells the story, that when he was quite young he never could bear to smoke ; and all attempts to do so left him on the floor of the room where he made the attempt. Recently, in middle life, he has tried it, and finds in it enjoyment, without apparent harm.

There is a story of a Greek philosopher, who vowed that he never would call in a doctor ; but when he was sick he sent for one, and when he was reproached for his inconsistency, he replied :

"But I am not the same person that I was."

To-day I cannot take coffee without immediate and perceptible injury ; ten years hence I may indulge in it freely without harm. Time was, perchance, when tea kept me awake the whole night long, but now it seems rather to dispose me to slumber.

You wonder that the cigar in which you formerly indulged without discomfort now excites a myriad un-pleasant and harassing symptoms ; that the glass of cider which once caused you intense headache, now serves as a tonic and appetizer.

To the rule that infants and children should abstain from stimulants and narcotics, there would appear to be some exceptions, of race, and climate, and condition; for in some countries babies use wine and beer, and little children, of both sexes, smoke cigars, and the query is not raised, whether they are or are not injured by so doing.

Their effects are greatly modified by disease.

There are forms of disease in which enormous quantities of stimulants and narcotics, of certain kinds, can be borne without difficulty ; and are, indeed, frequently curative.

In child-bed fever, for example, hundreds of grains of opium can be given daily, for a week or two at a time, to a patient who, in health, a very few grains would put into the sleep that knows no waking. By this bold dosing a disease that was formerly very fatal is now very frequently cured, leaving no unpleasant reminders.

In typhoid fever, quantities of alcohol can be taken which, if the patient were in his usual health, would keep him dead drunk, and leave him with racking headache and a muddled brain.

In certain chronic conditions of debility, also, great freedom of indulgence in punch, and whiskey, and wine, and beer, may be allowed without harm, and with great positive good.

There are, however, chronic diseases in which alcohol will not be tolerated, even in small doses.

Their effects are modified by habit.

The system can be educated to tolerance of almost any poisonous substance, provided the doses be

sufficiently small, and be given at proper intervals.
The amateur in tobacco smoking gazes with wonder,
if not admiration, on the man who takes his twentieth
cigar before retiring, but in a few years he may successfully rival him.

Of the effect of habit in causing tolerance of both
tobacco and alcohol, the whole German people are examples. Even in our climate they can smoke and drink
more freely than many other nationalities and with
less perceptible effects. It is amazing how the human
frame wears itself to the habitual use of stimulants and
narcotics.

Few poisons are more positive in their effects, or
more apt to be cumulative than strychnine, and yet it
is said that in very small quantities it is taken in India,
as arsenic is taken in Styria.

It is certain that in the East, where opium is produced, its habitual use appears to do much less evil
than with us. The masses would appear to be injured
by it far less than would be supposed; but in our climate, few indeed are they who can use opium for a long
time without harm. Among the Western nations there
are individuals who become so accustomed to opium that
they can drink laudanum almost like water, and yet attain good longevity.

Those who are beginning the habit of arsenic taking,
usually confine themselves to one quarter, or one half a
grain. They take the dose two or three times a week,
in the morning, on an empty stomach. This dose is
increased very gradually and with great caution, and
no bad symptoms appear.

If the individual who has thus formed the habit
leaves off, bad symptoms appear; loss of appetite,

burning feeling in the stomach, spasms in the throat, constipation and difficulty of respiration.

These symptoms, which indicate poisoning by arsenic, are only relieved by a return to arsenic eating. The effect of this habit among the inhabitants of Styria, is to improve the appearance of the skin and to impart color and beauty to the cheeks.

Those who indulge in this habit are not only healthy, but attain a good longevity.

They are all liable to be used to injurious excess, and to make slaves of those who indulge in them.

This statement is true of fermented and distilled liquors of all kinds, from the mildest wine up to the strongest brandy ; of all the vegetable stimulants and narcotics, from tea and chocolate to hashisch and opium ; and of the very powerful but less used anesthetics, as ether and chloroform.

Any of these substances injures when it is used in such a way as to produce a narcotizing instead of a stimulating effect. Narcotism is a degree of paralysis, and paralysis is always a disease, and injures the system in proportion to its intensity and continuance.

EFFECTS OF EXCESS IN TEA AND COFFEE.

Tea, coffee and chocolate are comparatively innocent beverages, and only in exceptional cases do they rapidly destroy the constitution, and yet there are thousands who cannot use them without injury even in their mildest preparations. They do not craze the brain like alcohol, or opium, or hashisch ; they do not directly and speedily ruin the manhood, and yet in not unfre-

quent instances tea and coffee, in the large and powerful doses in which they are employed in this country,
do cause not a little dyspepsia and frequent nervousness, and may slowly, and perhaps imperceptibly, un-
dermine the constitution, impoverish the nervous force,
and prepare the way for the inroads of all forms of nervous disease.

THE NERVOUS DIATHESIS.

There is a type of constitution which is much more
prevalent in this country than in Europe, which
is the result partly of our climate, and partly of
our habits and institutions, to which I have given
the name of the *nervous diathesis*. Those in whom
this diathesis exists, exhibit a tendency to disease
of the nervous system—hypochondria, nervous dyspepsia, sleeplessness, neuralgia, paralysis and nervous
debility.

This diathesis, like every other good or evil tendency
of the constitution, is strongly hereditary, and may run
in families for many generations, appearing in one
branch as epilepsy ; in another as insanity ; in another
as St. Vitus's dance ; in another as confirmed and per-
sistent melancholy ; in another as life-long dyspepsia
and constipation ; and so through the dark catalogue
of nervous diseases. Persons in whom this nervous
diathesis exists are peculiarly susceptible to the unpleasant and injurious effects of stimulants and narcotics, and are frequently unable to use habitually, even
the mildest forms without injury that is both sure and
perceptible. This diathesis, in its various degrees of
development, is exceedingly frequent in America ; and

is, to all appearance, increasing in **frequency** with the **increase** and **concentration** of **wealth, culture,** and luxury, and **the** general advance **of a** feverish civilization.

EFFECTS OF EXCESS IN TOBACCO ON THE NERVOUS DIATHESIS.

Tobacco is very apt to disagree with such **constitutions so markedly and so** rapidly that no physician **is needed to give the word of warning.**

I continually meet with patients who have found out, by observation **of** their **own** symptoms, that their favorite cigar, without **doubt, is doing** them injury

General nervousness, emaciation, sallowness, dyspepsia, sleeplessness, lethargy, **hypochondria,** actual paralysis—these are **some of the pitiful** evidences of over-indulgence **in tobacco in the forms** of smoking and chewing, that are patent to the **eye of** the observer, if not of the victim himself.

Other less marked signs of injury, that tobacco users may well look sharply for, are unsteadiness of the hand in writing, **or** in carrying any light object, as a cup, or **book,** or paper ; ugly dreams, bad appetite, constipation, fretfulness and impatience, headache, disturbance of vision or taste, and **impairment of** virility.

EFFECTS OF EXCESS IN ALCOHOLIC LIQUORS.

There is scarcely **a** nervous disease **known to** science that excess **in the** use of alcoholic liquors may not bring on or aggravate. Dyspepsia, general debility, neuralgia, insomnia, epilepsy, paralysis, of every form **and** type ; insanity in all **its** grades, as well as delirium tremens, may find in alcohol their exciting and their predisposing cause.

There is a disease that is well known in science, and is increasingly frequent. It is *chronic alcoholism*. It is a degeneracy of the brain, resulting from long excess in alcoholic liquors.

The leading symptoms of this disease are sleeplessness, nervousness, giddiness, headache, hallucination, debility, and difficulty of breathing.

DIPSOMANIA, (METHOMANIA—OINOMANIA.)

There is a special form of monomanic insanity that is peculiar to drunkards. It results from disease of the brain, through repeated and long continued alcoholic narcotism. Its leading symptom is utter want of control over the appetite for alcohol. Pledges of total abstinence, fear of disgrace, loss of social position, loss of friends and of family, apprehensions for this world or the next, all these considerations have no more weight against this disease than they would have against small pox or typhoid fever.

This disease has probably always been known, but it has greatly increased under our modern civilization, and especially in our country.

While it is difficult to name any type of nervous disease that the use of stimulants and narcotics may not in individual cases bring on or aggravate, yet to determine whether any individual case of disease is due to the tea, the coffee, the tobacco, which the patient has been wont to use, is almost always a question of exceeding difficulty.

ERRORS OF NEWSPAPER REPORTS.

Nearly all the newspaper reports of disease or death caused by the use of tobacco, and many of those attributed to alcohol even, are unreliable, and will not bear a critical examination. It is a very easy thing to assert, in any given case of paralysis, hypochondria, insanity, or other phase of nervous disorder, that the disease is the direct result of some narcotic or stimulant to which the person was addicted ; but our lives are so continually beset by such complex, diverse and opposing influences, both good and evil, that to trace the correct relation of causes and effects in our physical history, is a problem that is always difficult and usually impossible.

I have seen many cases of disease that I strongly suspected were directly and exclusively the result of indulgence in tea, or coffee, or tobacco ; but I see but few cases of disease where I feel myself justified in saying positively that I know it to be the result of such indulgence. In this, as in many other departments of science, positiveness of opinion is oftentimes in inverse proportion to knowledge, and he who is the most wise is also the most skeptical. The scientific man first smiles at the simplicity and enthusiasm of ignorance with which nearly all our periodicals and treatises on hygiene, and individuals in private conversation, declare for or against tobacco or alcohol, but afterwards shudder at the vast amount of evil that is thereby visited on society, and especially on the forming minds of youth, who are thereby trained not only to actual error concerning the nature and use of these agents, but what is of greater importance, to the habit of regarding statement for truth, and of neglecting the one and only method of finding truth, which is by looking at the facts and judiciously sifting the evidence, and without asking for an instant whether the result will tell for us or against us, or whether the apparent tendencies of that truth will be good or evil.

All over the land the cry has been lately heard—and its echoes are not yet dying away—that the marked increase of insanity of recent times is the result of the inordinate use of tobacco. Without expressing any opinion concerning the truth or falsity of this report, it is sufficient for our present purpose to remind those who raise this cry, that nearly all the forms of stimulants and narcotics so popular in Europe and America, except native wines, have come chiefly into use within the last few hundred years ; that this period has been characterized by an activity of brain unprecedented in the history of mankind ; which activity has been attended by a thousand strifes, anxieties, pressures, trials, as well as labors of which previous ages knew nothing.

I suspect that the worst effects of intemperance in the use of these substances, and especially in the use of alcoholic liquors, are least known and least noticed. It is the silent destruction of the nervous system ; the slow poisoning of the great centers of thought ; it is the transmission, by inheritance, of the evil from parent to child, from generation to generation, even more, perhaps, than the groans of the widow, the cry of the murdered man, or the tears of the orphan, that has made the temperance reform a necessity.

EFFECTS OF EXCESS IN OPIUM.

Even in the climate where it grows, opium, when used to great excess, and for a long time, destroys the constitution.

It increases thirst, lessens appetite, constipates the bowels, enfeebles the pulse, and exhausts the nervous system. In the last stages its direful effects are seen in the glassy eyes, the sallow features, and the unequal gait. Those in the East who give themselves wholly up to its seductions, are usually short lived. They are said to die on the average before forty.

Opio-mania is sometimes as truly a disease as dipsomania.

EFFECTS OF EXCESS IN HEMP.

On the Eastern nations, the effect of excess in hemp is to cause a state somewhat resembling catalepsy. The evil effects of over indulgence are less marked than those that follow excess in the use of tobacco or opium.

Instead of the very delightful enjoyment that some experience from a dose of hemp, there may be the extremes of agony. It is in this respect one of the most capricious of medicinal agents.

Experience of the author. I shall never forget the night of horror that I passed through after taking a not very large dose of this drug.

My object in the experiment was to see what effect it produced on the central nervous system, compared with opium, strychnine, and so forth. I settled the question so far as my own person was concerned, to my satisfaction, and for a while to my terror.

I took the dose just before retiring, about ten o'clock. About two o'clock I awoke with a strange and indescribable feeling in my head, and I at once knew that I had taken too much, and felt not a little alarmed.

I seemed to be in the midst of a great amphitheatre, the seats of which were filled with little devils, all incessantly bowing to me and grinning at my agony. I arose, went to the office of a physician near by, called him up, and tried to tell him my symptoms. Fully conscious of what I had done, I was yet unable to fully control my mental operations, and was unpleasantly conscious of this inability. The day following, the doctor informed me that I had not uttered a single connected sentence.

He induced me to go home and keep quiet, promising that all would be well. I dared not trust him ; my horror increased. The whole length of the spinal cord seemed on fire, and waves of heat rolled up and down my back, and through all the nerves of the body to the extremities.

I could have mapped out my whole nervous system by sensations of heat throughout my body.

I awoke the next morning as early as usual, and if anything, more bright and clear. I kept all my professional engagements for the day, and the only trace of the night of horror, was a slight pain in the back.

A similar over-dose of opium, of alcohol, of tobacco, or of almost any other stimulant and narcotic, would have been followed by a reaction.

The readiness with which it allows the system to rally from its influence is a striking peculiarity of hemp.

The delightful effects of hemp-eating, that have been

so often described—the visions of heaven, the double consciousness, the annihilation of time and space, and the glowing enjoyment—are experienced only by a portion of those who take it.

They can be better borne during or after meals, than on an empty stomach.

A very limited quantity of alcohol taken before breakfast may give rise to serious disturbance on one whom, after or during dinner, in the latter part of the day, several glasses do not affect.

Those who use tobacco and alcoholic liquors, very soon find out this fact 1 y experience. One of the worst things about the American custom of treating, is that it encourages a habit of taking liquor unaccompanied by food.

The custom of taking a smoke after dinner is founded on the observation that it does the least evil and the most good at that time. Raw spirits are particularly injurious when taken on an empty stomach.

Their effects are liable to be more or less complicated by adulteration and imitation.

Tea is adulterated with at least twenty-five different herbs, and with "lie tea," which is composed of the sweepings of the warehouses.

Green tea is sometimes colored with black lead ; also Prussian blue and indigo or gypsum may be used. Catechu, gum and starch are sometimes added to tea.

Coffee is adulterated with carrots, peas and beans, corn, burnt sugar, and especially with chicory. Chicory is itself adulterated with Venetian red, and Venetian red is adulterated with brick dust. Chicory is said to be adulterated with seeds, acorns and the baked liver of the horse.

Cocoa is adulterated with arrow root, starch, sugar, and red and brown earths.

Tobacco is adulterated with at least sixteen different substances : molasses, sugar, honey, rhubarb, the birch, the walnut, mosses, bran, beet-root, drugs, liquorice, rosin, yellow ochre, fuller's earth, sand, saltpetre and common salt.

In Persia wine was adulterated with poppy heads, and in ancient Palestine with frankincense. Wine is now adulterated in numberless ways, and so far as counterfeiting the taste goes with amazing success. A gentleman engaged in the sale of California wines told me that the choicest varieties could be counterfeited so accurately that he could not himself distinguish the genuine from the false.

With alcohol as a base and a proper variety of chemicals, almost any kind of wine can be imitated with more or less success. Pure champagne is quite rarely seen, common cider and even petroleum are said to be the basis from which some of it is manufactured. Even those who pay the highest price are not sure to obtain the pure article. The term "chain lightning" that has been offered to these mixtures is entirely appropriate.

Among the more prominent substances that are added to alcohol or cider to imitate wines and ardent spirits may be mentioned catechu, burnt sugar, grains of paradise, logwood, bitter almond, honey, mustard, orris root, cassia, kino acetic acid, acetic ether, tannic acid, oil of cognac and œnanthic ether, pepper, red sanders, prunes, nitric ether and pellitory. There is no evidence that strychine is much used to adulterate **either fermented** or distilled liquors. Nearly all the brandy-champagne and port wines used in this country are either adulterated or counterfeited, and sherry is more frequently impure than pure.

Rhine wine is probably **less** adulterated **and counterfeited** than the other wines, and very **fortunately is** the best adapted for habitual use.

Of the ardent spirits whiskey can be obtained **in** greater comparative purity than any other form.

Claret is made, it is said, in large quantities by allowing water to soak through shavings, and afterwards adding water, tartaric acid and alcohol.

Even pure cider is extremely hard to obtain in our large cities. I have been told by a merchant in the wholesale grocery business that he sells quantities **of** sugar to manufacturers of cider ; sugar, tartaric acid, water, coloring **matter,** and perhaps a small quantity **of** the real article, make an imitation that is **profitable to** the manufacturer and not very injurious to the consumer ; but very disappointing to him who seeks for the fermented juice of the apple. Honey, **whiskey,** alum are all employed in the manufacture of cider.

Beer is adulterated with cocculus indicus, sweet flag and thorn apple, grains of paradise, lime, **soda, salt,** alum, pepper, capsicum, gentian, quassia, **the marsh** ledum and **even** tobacco **leaves.** Beer made from **rice** is adulterated **with** red and black pepper **and** onions. While adulteration is, as all know, the rule rather than the **exception** in this country, yet the *injurious* character of the adulterations of liquors is *by no means so great as has been commonly supposed.* All of them contain alcohol as a **basis, and most** the other ingredients are either harmless or are in such small quantities that they can exert but little evil effect on the system. Some of the adulterations of our ordinary food, and especially of our candies, are far more harmful than those which are used in imitating or adulterating liquors.*

---

* See my work on " *Eating and Drinking* " of this series.

CHAPTER III.

ALREADY we have given the history of stimulants and narcotics in general, and we have taken a general survey of their effects as variously modified.

We have seen that all of them are liable to be used to excess. The slaves to these substances receive different names in different countries, and according to the special substance in which they indulge. Thus, in India the slave of opium is called a "thierak ;" in South America the Indian who becomes a slave to coca is disgraced as a "coquero ;" in Europe or North America coca is not used at all, and opium is the curse only of a comparatively limited number; but intemperance in alcoholic liquors is one of the greatest evils of civilization, and especially of Great Britain and the United States.

Coleridge used to say that if he lived by the sea-shore he would preach fifty-two sermons a year against the wreckers. On the same principle the philosophy and treatment of intemperance is worthy of a special chapter in any work on stimulants and narcotics that is designed for an American or English audience.

What is Intemperance ? Why is it that man, the highest in the animal world, makes himself the lowest

by indulgence in stimulants and narcotics? Why is it
that man, with his powerful will holding him back from
wrong, and his rich and varied moral nature lifting
him above the temptations of passion as no other animal
is favored, yet becomes the worst of animals in the
presence of opium or alcohol?

THE LOWER ANIMALS NOT INTEMPERATE.

Other passions the lower animals indulge more reck-
lessly, more grossly than man, but not this. Give them
access to opium and alcohol, and, as we have seen, they
care not to become drunk; they are not even moderate
drinkers. Who ever heard of a cow, or a horse, or a
sheep, or dog, or cat, or a hog, going down to a drunkard's grave?

These animals, overeat; they shorten their lives,
and greatly diminish their happiness by their gluttony
and self-indulgence. They acknowledge no moral law,
no conscience, they do what they like, as the phrase is,
and care not for yesterday or for to-morrow.

What maketh us to differ? *Mainly our nervous system.* Man has a larger, fuller, richer brain than the
lower animals, and stimulants and narcotics chiefly affect the brain, therefore man craves for them, finds rest
and negative food and pleasure in them, and thus often
becomes their slave. The horse does not care for alcohol for the same reason that it does not care for philosophy, because its brain is not able to appreciate it.*

*Intemperance in the use of stimulants and narcotics is
quite frequently a disease—a symptom of cerebral disorder*

* Monkeys sometimes acquire a taste for tea, coffee, and alcoholic liquors.
They have been known to get drunk, and Darwin says that he has seen them
smoke tobacco with pleasure.

of some kind, induced by the use of other substances, **or** *inherited.*—**Like** other diseases **of** the brain, it may be inherited ; **may skip** over two **or** three generations, and break out in a **family that** supposed it had long been delivered **from its** presence.

Like other chronic nervous diseases, **it is very obstinate and sometimes utterly incurable.** Like **other chronic brain diseases, it needs both** physical and metaphysical **medical treatment—medicine for** the congested **or** exhausted **brain, as well as rest,** relaxation, advice, care, **watchfulness, exhortation, and in some cases compulsion. Moral or** metaphysical treatment **alone will** not avail **to cure** it, usually, any more than **it will avail** to cure **epilepsy,** or neuralgia, or paralysis, **or insanity.**

Intemperance as a vice.

I have said that intemperance was **oftentimes a disease ; but it is** not always so. It appears **in two forms ; one is a** symptom **of a** diseased state **of the brain ;** the **other is a** symptom **of a** viciously **organized, but entirely** healthy brain. *The one* **I call** *intemperance from disease, the* other, *intemperance from* **habit,** *through bad organization and surroundings.* Intemperance from **disease is the** form which **is most frequently** found **among** the intellectual **and the cultivated ;** intemperance **from a** bad organization **is the form** which is most frequently found among **the** ignorant and degraded, **and** among the so-called criminal classes ; and yet both forms appear in all grades of life. Some men **are** born to intemperance, **just as** the sparks **are prone to** fly upward. Crime **of** all kinds is to **a certain** extent organic, and many **of** our criminals are often **subjected** to their own evil organizations, even more than to the laws. . Either from an excess of some qualities, or from a deficiency of others, or from both causes, their brains, though perfectly healthy, are not modeled after the type of good men, but rather of bad men, and it is as natural for them to get drunk, or to stupefy themselves with opium or tobacco, as it is for other and better formed natures to study philosophy, to write poetry, to succor the destitute, or to fall on their knees in prayer. The drunkard in the gutter, and philanthropist who lifts him out, may be both acting in obedience to organization, for which they deserve but little praise or blame.

There are those, especially among the lower classes, who get drunk from a spirit, which that very distinguished philosopher, Artemus Ward, has hammered down into a definition of two words—"*pure cussedness.*"

They love ugliness for its own sake.

Both forms of intemperance—that which is a vice, and that which is a disease—like all other diseases and vices, are greatly modified and varied by race, climate, religion, education, and other external conditions.

How shall intemperance be treated?

Before attempting to answer this question, it will be necessary to take a brief survey of the history and present state of this evil in all parts of the world, in order to learn in what way it is modified by race, climate, sex, education, and external conditions, and by the different forms of alcoholic liquors used ; and also to learn what success has attended the various efforts to cure this evil.

Information on this subject I have obtained from a wide variety of sources : from the literature of stimulants and narcotics ; from criminal and police statistics ;

from personal observation of the customs of different nationalities, and in different countries ; from works of travel, and from conversation with foreigners and travellers.

It is not a little remarkable that the first systematic attempt to gather facts regarding the subject of stimulants and narcotics, all over the globe, should not have been made until the last year. With all the excitements to which the liquor question has given rise ; amid all the battles that have been waged between the hosts of temperance and anti-temperance, between prohibitionists and anti-prohibitionists, between teetotallers and moderate drinkers, it seems not to have occurred to any of the parties engaged, that a carefully arranged collection of *facts* relating to the use and abuse of intoxicating substances might, perhaps, shed light on the points of controversy, and perhaps settle the discussion.

If reason could have been allowed to speak, or if her voice had not been drowned in the noisy rabble of the passions, this method of reaching the truth, on a most complex theme, would long ago have been suggested and carried out, and the country might have been saved not a little of rancor and evil speaking ; and the temperance reform would have been established on a surer and firmer basis.

ENGLAND.—In England nearly all classes drink some form of alcoholic liquors. The upper and middle classes use imported wines of all kinds on their tables, and the poorer classes indulge in beer, whiskey, and cheap and adulterated wines.

In spite of the agitation on the subject of temperance,

the number of total abstainers in England in any rank of society is very small. In this respect England is greatly different from the United States, where among the ruling orders approximate or absolute teetotalism is the rule. Although the custom of "treating" and drinking at public bars is not as prevalent among the wealthy and cultivated classes in England as in America, yet there are very few families who do not regularly have wine and beer on their tables.

Among the peasantry, who constitute the great majority of the population, beer and cheap wines are not only used but are terribly abused ; they drink their beer not only by the glass but by the pint and the quart; not only at meals, but at all hours of the day ; not only at home, but in the ginshops and lunch bars that abound on every hand ; at the theatres, in the intervals between the acts, beer is brought around, sold and drunk with a freedom that astonishes an American, not only by adults but by children, and even infants. Very often I have seen mothers pour abominably bad beer down the throats of babies in their arms. At public places it is the custom for a family to buy a quart mug full at a time and all drink from it as much as they wish.

The climate of England, like that of Scotland, Ireland and the countries of the interior in the same latitude, is so very moist that very large quantities of liquor can be taken without apparent discomfort. Probably there is no portion of the world where the stronger alcoholic liquors are borne as well and as freely used by all classes as in Great Britain ; and yet that the lower classes especially, and even the cultivated, are injured by the vast quantities of alcohol which in various ways they daily consume, even where **they do** not fall into **habits of** drunkenness, there **can be no doubt.** Love of **good eating and** drinking **is a strong peculiar** passion of the **race** to which the English belong.

INTEMPERANCE IN ENGLAND.

England is one of the most intemperate **countries in the world.** Formerly, as in Scotland, **all classes from** peers to peasants indulged in drunkenness, and the occasional habit of intoxication was not disreputable. **The** custom for ladies to retire at or near the close of dinner while the gentlemen remain, **and which is still preserved in** English society, originated in **necessity, since the dinners were pretty sure to be followed by drunkenness.** The last half century and even the past quarter **of a century has witnessed** a vast reform in **this matter,** and among **the middle** and upper classes drunkenness is no **longer honored, and** the reputation **for a habit of** excess in drinking is **a** bar to social **advancement.** Although ladies retire toward **the** close **of the feast** as of old, yet **the** necessity **in which** the **custom** originated has long **since** pased away. Men remain at the table as formerly **to** chat, and **smoke,** and **drink,** but instead **of** port and brandy **they use** mostly **the** lighter wines, claret and hock, **and in less** quantity.

Among **the poor and** ignorant a different **story must** be told ; they are **more** intemperate than formerly, and in spite **of** the efforts of temperance men **they are** growing **worse** and worse. There probably never was a time when **the** English peasantry **were so** intemperate as now.

By the **Police Report of London,** which I obtained at **the central office, I found that out of** 63,000 arrests in 1862, 17,000 were drunk or disorderly, or both, and probably many of the other offences were directly caused by drunkenness.

By the Police Report of Liverpool, one of worst places in the world, out of 26,000 arrests in 1868, 16,000 were intoxicated.*

TEMPERANCE REFORM IN ENGLAND.

The first temperance society in England was formed in 1834, at Lancashire ; September 24th of the same year, the total abstinence platform was adopted at Manchester. The word teetotaller is derived from the fact that one of the members, on signing the pledge on this occasion, stammered very badly over the words total abstinence.

In 1839 Father Matthew began his wonderful mission, and obtained thousands of signers to the pledge.

Total abstinence has never been as popular in England as in the United States, and very little has yet been attempted by legislation.

Sir Edward Coke laid down the maxim that drunkenness aggravated a crime.

In the time of James I., drunkenness in England was punished by a fine, or by public exposure in the stocks. In 1828 the law against drunkenness was repealed. Excise was introduced into England in 1643.

Recently, on account of the great increase of intemperance among the lower orders, the question of prohibition has been seriously discussed.

VALUE OF STATISTICS IN SCIENCE.

I know how hard it is to learn what fact is; how statistics may lead to error as well as truth; how many general

---

* I here give the round numbers.

considerations there are, and deflecting influences and modifying conditions, and elements, and factors, that subtract from the value of statistics ; how easy it is for them to master us, instead of our mastering them, and those who have done the most in gathering statistics are best aware that, like all good things, they may be perverted.

But in spite of all, statistics are the keys by which we unlock the gate of truth, and they have solved problems which, without them, would be unsolvable. They are pouring light on questions that were dark as night. They are gaining in popularity every year and day with the best scientists of the world.

SCOTLAND.—In reply to the question, "What is the chief intoxicating liquor used in Scotland?" Dr. Christison, of Edinburgh, writes : " The foremost is whiskey; the next is whiskey ; the third is still whiskey."

Scotland has long been famous for its intemperance.

Fifty years ago, and even less, the drinking customs of Scotland were, perhaps, the worst in the world. Everything was associated with drinking. Funerals and weddings, minister's meetings and church going, and every important or unimportant public or private act needed whiskey to make it successful. To get drunk, even beastly drunk, to drop under the table after dinner, was, in many circles, more of an honor than a disgrace.

For a time, the extreme religiousness of the Scotch churches seemed to have no effect to diminish this awful crime.

About fifty years ago the temperance reform began, and among the better classes its progress was very rapid ; and among these classes drunkenness has ceased to be an honor and has become a disgrace.

Among the peasantry, the reverse has occurred. They are more intemperate than formerly ; and all the efforts of the temperance organizations among these classes have been of little permanent value.

A great proportion of the crime of Scotland is in some way connected with intemperance.

IRELAND.—That Ireland is a great center of intemperance is too well known. Statistics show that the average annual expense of Ireland for liquors of all kinds, including wines and beer, is about *fifty dollars for each family.* It should be considered that liquors are not as cheap in Ireland as in some of the Southern countries, where the grape grows, and that, therefore, fifty dollars' worth does not represent so large an amount of liquor as it would in some other countries.

It should also be considered that the Irish are poor ; that wages are low and rents are high ; and that fifty dollars is a very large sum for the great mass of the population.

The poorer classes of the Irish drink ale, porter, and whiskey ; the wealthy and middle orders, brandy and wines.

It is stated, also, that in some of the Northern towns, as Draperstown and Maghera, " æther has been used to a very considerable extent."

There is no question that the greater portion of the gross crimes of Ireland are caused by drunkenness ; and there is no more question that the health of the lower orders is most seriously affected by their drinking habits.

Russia.—The population of Russia is very largely composed of emancipated serfs, laborers, and soldiers.

These are coarse, grossly ignorant and superstitious. With this class the one great drink is "*vodki*," which is very much like whiskey. It is of different colors and is variously spiced and flavored. It intoxicates like whiskey, and the poorer classes are grossly intemperate.

The holidays in Russia are more numerous than in any other country, and on these days and on Sabbaths the peasantry may frequently be seen staggering about or lying dead drunk.

A large proportion of the crime of Russia comes from intemperance—probably one half or three fourths.

By the higher classes imported wines, beers, ales, brandy, etc. are used with their meals, and especially at dinner and breakfast, but intemperance is quite rare among them. All this is true of Siberia.

Local temperance societies have been formed in some parts of Russia.

Sweden and Norway.—In Sweden and Norway, as in all the northern countries of Europe, the stronger forms of alcoholic liquors are used more or less by all classes, and intemperance is a frequent vice. In Norway the "Bairisch" beer of Germany is much used in the cities, and whiskey among the farmers in the country, especially on and about Christmas.

Of all countries, Sweden has enacted the severest laws against intoxication. Drunkenness itself is a crime even when it does not lead to disorder. He who is seen drunk is fined for the first offence $3; for the second $6; for the fourth $20; or is publicly exposed in church on the Sabbath; for the fifth offence is

confined in the House of Correction ; for the sixth offence is condemned to twelve months hard labor. It
is not allowed to sell liquors to students, workmen,
servants, apprentices and private soldiers—that is, to
the classes that are most likely to become intoxicated.
Of the fines, half go to charity and half to the informer.
These laws are read several times a year from the pulpit, and every tavern keeper is obliged to keep them
hung up in the principal rooms of his house.

DENMARK.—Alcoholic liquors are used very liberally
in Denmark, and their consumption has certainly increased during the past quarter of a century.

Beer of a mild character is universal. Brandy, distilled from barley or potatoes, is manufactured in very
large quantities and is much used. The higher classes
use more or less of foreign wines.

The people appear to be very little injured in health
or morals by these liquors.

Intemperance is rare, and, consequently, the crimes
that result from intemperance are rare.

It is a fact of considerable interest that while the
consumption of alcoholic liquors has *increased* of late
years, the amount of visible intoxication has *diminished.*

This apparent paradox may be explained by the fact
that during the time in which the consumption of
liquors has been on the increase, the Danes have been
rising in the world, so that their educational, political,
and economic state is greatly improved.

The average standard of intelligence is very high.

Brandy of a very strong kind is used by the laboring
classes with every meal, and they have four or five

meals a day.   The subject of temperance has been little thought of in Denmark.

It has been estimated that twenty gallons of beer annually, for each head of the population, are consumed in Denmark.   The beer is very cheap, being sold for one, two, or three cents a bottle.

Between four and five gallons of wine to each individual are annually used.

The Danes, it should be remarked, are wonderfully strong, quiet and orderly ; but are not very active.

Austria.—The great drinks of Austria are beers and wines, and the relative consumption of beers appears to be on the increase.   Brandy is, however, considerably used.

It has been estimated that in Austria and Hungary, containing a population of about 35,000,000, the average annual expenditure for wines, beers, and brandy for each person is eight times as much as they spend for iron.

The Galicians, who use brandy to excess, seem to have degenerated, and it has been thought that their degeneracy is the result of their intemperance.   In Bohemia, Moravia, where beer is the principal drink, and in Hungary, where wine is most abundant, there is much less intoxication than in those districts where corn brandy is the staple drink.

Local temperance societies have been formed in some portions of Austria, particularly in the Galician provinces, but they have accomplished very little.

Prussia.—The principal alcoholic drink of Prussia is beer, " Vien," and " Bairisch " beer, which is used everywhere and in enormous quantities by both sexes and by

old and young.   Wines of various kinds, and especially
the Rhine wines are used ; but wine is not as cheap as
in France and Switzerland, and therefore, beer, to a
considerable extent, takes its place, especially among the
lower classes.   A man or woman of any order of society
in Prussia who totally abstains from wine and beer
would be very hard to find.

The Prussians also use brandy distilled from rye or
potatoes, and "Schnapp," which is the brandy distilled
with sugar.   The French liquors, absinthe, etc., are also
used in Prussia.

Very few in Prussia are injured apparently by their
wines or beer.

The amount of intoxication is small, and is the result of the indulgence in schnapps or the liquors.

The number of those among the educated classes who
become victims to intemperance is very limited ; but
instances now and then occur.

I searched the Police Records of Berlin and found
that, in one year, out of 32,254 arrests, only 733 were
for drunkenness.

RHINE PROVINCES.—The fact that wine is produced so
abundantly on the Rhine might, at first, lead us to suppose that it would be very cheap in the Rhine provinces, and would be the exclusive drink of the people,
as native wines are in other countries.

This is found not to be the case.   Rhine wine is not
very cheap along the Rhine ; and the better qualities
are quite expensive, even under the shadow of the vine-
yards in which they are produced.   At all the hotels
along the Rhine, and on the boats, high prices are

asked for the wines that are made from grapes that grow on the banks of the river.

The poorer classes drink beer and brandy ; and a kind of very cheap wine or cider, called "*tietz*," is considerably used. This latter is quite intoxicating.

The schnapps cause considerable intemperance among the lower orders ; and it has been estimated that "seventy-five per cent of the imprisoned became criminals by brandy."

Müller, the director of the Central Institution, at Cologne, says : "I am convinced that a great number of prisoners, if they had the opportunity, would rather stretch out their hands for a glass of schnapps than a piece of coal."

Most of the vagabondage of that country is attributed to indulgence in the *tietz* or schnapps.

It is said that in Frankfort the custom of drinking in dram shops has almost disappeared, during the last twenty years ; that the popularization of coffee and the improvement in the beers here, caused the stronger liquors and the wretched cider to be less used ; and consequently intemperance has diminished, although it still exists among the lower orders.

Among the intelligent classes of the Rhine provinces intemperance is very rare indeed.

NETHERLANDS.—Great injury is believed to have resulted to the inhabitants of Netherlands by the use of gin, which is "the beverage of the people." A society has been formed called the "*Netherland Society for the abolition of strong drink.*"

In the opinion of the secretary of this society, "the number of gin drinkers has considerably decreased ;

and the use of that beverage by the higher and middle classes is considered indecent."

Much of the crime is caused by intemperance.

SWITZERLAND.—In Switzerland, strange to say, there appears to be much more of intemperance than in France, Italy, or Germany. In Berne, a temperance society has been formed, the members of which abstain from all distilled liquors, but not, I believe, from wine and beer.

Laws have been made to increase the tax on spirituous liquors, and diminish the tax on beer and wine, in the hopes that the people might thereby be led to abandon the use of spirituous liquors, and confine themselves to milder beverages.

Wines and beers are largely consumed by all classes ; "schnapps," a "species of brandy, distilled from potatoes, or from the pulp of grapes after the wine has been pressed out," are used by the peasantry.

The use of schnapps has increased of late ; and this increase has been accounted for by the lack of good food among the poorer classes, who are obliged to live almost exclusively on potatoes.

Whatever intemperance there is in Switzerland appears to be due to the use of schnapps, and not to the use of wines.

ITALY.— The Italians use but little strong liquors. Native and foreign wines have been freely consumed, and no very serious injury seems to have resulted. They are used habitually, daily, and from the earliest years ; but mostly as a means of quenching thirst, and

sustaining the system, and not for the purpose of exciting the nervous system.

Recently, on account of the higher prices of wines,
spirituous liquors of a bad quality—rum, brandy, whiskey, absinthe—are beginning to be used ; and the effects
of these are sometimes very pernicious.

The amount of intoxication in Italy, however, is very
slight, even among the lower classes ; and comparatively
little of the crimes of the native population are attributable to the abuse of liquors.

The subject of temperance has, in Italy, never excited
any general interest.

FRANCE.—France is the land of the vine, and by all
classes wine is as universally used as bread.

The peasantry very largely depend on it for food, and
with vast numbers bread and wine are the main sup-
porters of life and toil. The wines used are chiefly
those made in France, and of these there are several
hundred varieties.

It is just to say that there are more in the United
States who do not use tea or coffee, than of Frenchmen
in France who do not daily use wine.

Besides wines, beer, brandy, whiskey, gin and absinthe are used especially in the cities, but not usually in
large quantities. The Frenchman sips his *liqueurs* from
very *small glasses*, like most of the southern people of
Europe, and takes much less at a time than the Englishman or American.

The chief curse of France is *absinthe*, which is made
from wormwood, which is fearful in its effects when
used in excess, and which unfortunately has grown in
popularity in recent times. This drink induces habits

of intoxication, and frequently destroys the constitution. Intemperance, though far less frequent than in England and the United States, is yet by no means unknown. What there is of it, is mostly confined to the lowest orders, and is increased in times of great excitement, like the late war.

There is reason for the belief that intemperance has increased in France in recent times, but the subject of temperance has never been formally agitated.

The French, as a people, like the Italians, and Spaniards, and Greeks, are but little inclined to intemperance, even when exposed to the extreme of temptation.

The published records of the police office gave me no information, for drunkenness is not regarded as a crime, and when any of the laboring classes are found drunk, they are sent or helped home, and nothing more is done with them.

Drunkenness rarely leads to great disorder or to crime in Paris and in France generally, or at least very much less frequently than in Great Britain and America.

SPAIN.—The alcoholic liquors chiefly used in Spain are cherry wine, burgundy and *aguardiente*, the " whiskey of Spain."

The effect of drinking these articles seems to be to make the people indolent, or, at least, to increase their native indolence.

They drink to excess, but an indulgence seems to make them good-natured rather than quarrelsome.

Very little of the crime of the country, and the amount of crime is very great, appears to be caused by intoxication.

TURKEY.—The Mohammedams are, by their religion, forbidden the use of wine, and habit has made abstinence with them a second nature.

Wines, native and foreign, are cheap in Turkey, and are freely used by foreigners and Christians, and are not believed to do harm. *Rakée* or *mastica*—that is rum flavored with mastic and brandy, are the principal spirituous liquors in use, and in these the foreign sailors indulge most frightfully.

JAPAN.—*Saki*, the chief intoxicating drink of Japan, is obtained by distillation of rice. There are various qualities of saki, some of which are as weak as ale or cider, while others are quite strong. Saki is used by all classes, but is drunk in very small cups, holding not more than an ounce.

The Japanese are very susceptible to the effects of the saki, and are flushed and excited by quantities that on a European would have no effect whatever.

They not unfrequently become good-naturedly drunk, especially on festal days, but this slight intoxication rarely leads to crime.

The women of Japan are accustomed to drink, but in a very quiet way, and very rarely to intoxication.

CEYLON.—The chief liquors of Ceylon are "*toddy*," made from the flowers and stems of the palm, and "arrack," prepared by distillation of toddy.

The toddy contains sugar, and is quite agreeable if taken before it has been long fermenting. Both arrack and toddy are very cheap and are used with great freedom.

There would appear to be more intemperance here

than in any other Eastern island. It is declared by those best qualified to judge, that the use of arrack—which is about as strong as whiskey, causes considerable intemperance and crime as well as disease. The government has let out this privilege of selling arrack wholesale to tavern-keepers, and as a result intemperance has increased.

SYRIA.—There is scarcely any intemperance among the inhabitants of Syria, and no evil effects of any kind can be attributed to their use of liquors. The native wine is used, but very rarely, if ever, to intoxication. The "*saki*" or "*rakia*," which is distilled from the wine is very intoxicating; but it is used with caution. The inhabitants do not seem to desire the exciting effect of liquors that are so fascinating to the Anglo-Saxons. They use them to quench thirst and sustain life.

Foreign brandies and whiskeys have been introduced, but are chiefly used by foreigners, who sometimes become intoxicated.

Dr. Thomson, a resident of Syria for about thirty-five years, states that he never saw a drunken man during the larger part of the time.

CHINA.—The Chinese drink "samshoo," a fermented drink made from rice; but they are not, as a people, intemperate. Their greatest vice is opium eating.

INDIA.—The Hindoos use a wretched fermented drink made from rice, and on this some of the lower orders become intemperate. The curse of India and of China is opium, which is used almost as freely as tobacco is used in Europe and America.

The English residents in India carry with them the drinking customs of England ; but it is found by experience that they must use a very far less quantity of alcohol than in England. *Extremes of heat like extremes of cold do not allow liberal indulgence in any form of alcoholic liquors ; they are best borne in climates that are temperate and moist.*

The British soldiers have introduced habits of intemperance into India that are said to have extended among the native population.

TENERIFFE.—The inhabitants of Teneriffe use wines, and rum, and gin.

Formerly wines were manufactured on the island ; but since 1845, when the vines were destroyed, the people have depended mainly on importation. Intemperance, formerly very rare, has somewhat increased of late years, though at the present time it is not prominent.

MALTA.—No wine is made in Malta ; but the inhabitants use red and white wines imported from Sicily. Brandy is also in use—especially among the poorer classes.

The wealthier classes use the principal wines of Europe. The foreigners—mostly English—use the same drinks as at home, and their example has been to cause increase of the habits of intemperance among the natives, who were formerly very temperate.

AZORES.—The inhabitants of Fayal are very quiet, peaceable and inoffensive.

The native wine is most used, but not so freely as formerly, because it is more expensive.

Since "the almost entire destruction of the vines in 1855," rum has been introduced, and to a certain extent has taken the place of wine, and yet there is but little of intoxication.

MADEIRA.—There is very little intoxication in Madeira.

Wines and cane brandy are the chief drinks, and are used at meals and between meals by all classes. The brandy is very strong.

SANDWICH ISLANDS.—The government of the Islands forbids the distillation of spirits; and the sale of intoxicating liquors to the natives is illegal.

Illicit distillation is, however, carried on to a considerable extent. The native spirits are made from the "Ti" root, and the "Ava" root. It has the strength of whiskey.

It would appear that prohibition had at one time been more successful here than in any other part of the world. Liquors of various kinds are imported for the use of the foreign white population.

GREECE.—The native wines of Greece are very pure and very cheap. The resinated wine, which is prepared by placing resin in grape juice and allowing it to ferment, sells for about three cents a bottle.

It is the universal drink of the people. It is taken with their bread and olives, and is regarded as an article of food. Both sexes, young and old, even babies drink it.

No injury of any kind appears to ensue from its use, and it is regarded as beneficial in every way.

The only form of spirituous liquor used by the Greeks is *rakée,* which is very strong ; but this is drunk by comparatively few. The rakée is made of the lees of wine and figs.

On the whole the Greeks are among the most temperate people in the world. Even when they do occasionally become intoxicated they do not fight, or kill, or strike, but simply quarrel with words.

Delirium tremens is very infrequent in Athens ; and of forty-two cases reported in ten years, sixteen were foreigners, although foreigners constitute only about one per cent of the population.

E GYPT.—The Egyptians drink arrack, champagne and wines.

Drunkenness is very rare in Egypt, as in other Mohammedan countries. The little intemperance that is seen is mostly confined to the foreign population. Among the *African Tribes,* palm wine causes much intemperance.

ZANZIBAR.—The negroes of Zanzibar drink cocoa-nut rum, and the resident Arabs use gin and brandy.

The religion is Mohammedan, and therefore drunkenness is very rare, except among the sailors in port.

WEST INDIES.—A drink much used in Hayti is "Tafi" or cane spirit. The natives are not free drinkers even of this article, which is constantly used, and sometimes to excess, by the English and American sailors.

Many of the natives of Hayti seem to have no desire to drink.

In St. Croix the famous Santa Cruz rum, as well as wines, brandy, and malt liquors are in use. The lower classes use rum manufactured on the island; and the higher classes, wines, and brandy, and beers, which are imported.

There is comparatively little crime of any kind, and that little is not attributable to intoxication. Among the higher classes the use of stimulants is universal; but it does not lead to gross intemperance.

In Trinidad the "aguardiente," distilled from molasses, is the main drink of the lower orders. It is very cheap, and is used for bathing.

Claret wine is imported, however, and by the male population is habitually used.

Drunkenness is looked upon as a most revolting crime, and is very rare.

Opium-eating among the coolies is very frequent, and leads to much evil.

NICARAGUA.—The native rum of the tropical countries, made from sugar cane, is used abundantly by the poor, and does not work serious injury.

Among the ruling orders, imported liquors of all kinds, mostly adulterated, are used; and not unfrequently to excess.

BRAZIL.—The Brazilians drink with utter freedom and without apparent injury, the "cachaça," made from the sugar cane.

The negroes sometimes become intoxicated; and the Indians, also, especially on festal occasions.

Drunkenness is held in very great reproach, and is unfrequent.

In Brazil and other South American countries, ale, and porter, and wines are imported from Europe.

Most of the crime caused by intemperance is committed by foreigners.

PERU.—Into Peru all the principal wines and liquors of Europe are imported, and besides native brandies also " chicha," made from maize, and similar in taste and character to the beer in our whiskey distilleries after fermentation, is much used.

The native wines are considerably drunk by all classes.

The health of the Peruvians is excellent. Many of them reach great age, and intoxication is very little seen or thought of.

PANAMA AND DARIEN.—The residents of the Isthmus of Darien use " anisette, cocoa-nut milk, wine from the wine palm, a drink made from bananas and plantains, and a milky-looking liquid made only by the Sassardi-Morti Indians."

The anisette is very powerful and very injurious.

There is almost no intemperance in Darien or Panama.

On other places of the isthmus, where the Indians are more advanced in civilization, there is much more of intoxication, but yet much less than in the United States.

MEXICO.—The notorious drink of the Mexicans is the fermented juice of the agave plant.

Foreign wines and stronger liquors are also imported and used more or less by the foreign and wealthy classes.

The Mexicans are not an intemperate people, are not disposed to become sots, and in this respect seem to resemble the French, Spanish, and the South Americans generally.

CANADA.—The drinking customs in Canada do not much vary from those of France and England. All classes, high, low and middle, drink some one or many forms of alcoholic liquors. The number of total abstainers is much less than in the States. At dinner their use of these substances is almost habitual, as in England.

The liquors are purer and cheaper than in the United States, but it is observed that it is not possible to drink as large quantities as in England and Scotland.

The amount of intemperance among the poor and ignorant classes is terrible, just as in England and the United States.

There is no doubt that the greater portion of crime in Canada, is in some way connected with the use of alcoholic liquors.

UNITED STATES.—The drinking customs of the United States are as diverse as are the nationalities that compose it. Here are to be obtained nearly all the forms of alcoholic liquors in use in Europe, besides some that are peculiar to itself. To our ports are brought West India rum, the brandies, gins, champagnes, and wines of the continent ; the porter and ale of Great Britain ; while whiskey, and cider, and lager, and other beers, and wines, also, in some variety, are manufactured at home.

All these substances are variously used and abused, according to the race, the social position, and the individual temperament. The general tendency for recent emigrants is to continue the customs to which they were brought up, in the districts from which they came, with this difference, that they are, as a rule, obliged to use less quantities of alcohol, on account of the peculiarities of our climate, of which I have previously spoken.

Here, as in Canada, those who, in Europe, have been wont to indulge in enormous amounts of strong whiskey, or gin, or brandy, without perceptible injury, find that they must retrench or suffer. The Germans, who use lager beer, probably appreciate this effect of the climate less than any other class, for the two-fold reason that they are so little impressible, and because the proportion of alcohol in lager beer is so very small.

Prior to the temperance reformation in the early part of this century, the custom of drinking freely strong liquors was absolutely universal. Strong spirits were used more liberally than tea and coffee at the present time ; and the relations of habits of drinking to morality were never for a moment considered. Between the church and the world there was, in this matter of drinking, no distinction ; and drunkenness itself was not a disgrace. The early settlers of New England were descended from the Anglo-Saxons, a race among whom the ability to hold a large quantity of liquor was regarded as a heroic virtue.

During the last half century, partly as a result of the special efforts made by the temperance reformers ; partly as an accompaniment of the general elevation and improved culture of society, not only the vice of drunkenness, but the habit of moderate drinking itself has ceased to be universal, and is confined to certain classes or orders of American society. At the present time the great middle class of American society—the solid yeomanry, the majority of those in the leading professions of law, medicine, teaching, and the ministry, and the best of our merchants—are either total abstainers or drink very sparingly and irregularly.

Cider is used pretty freely in some farming districts of the North ; but from the tables of the greater portion the substantial part of our population alcoholic stimulants, of all forms, are absolutely banished.

There are thousands of families who would as soon think of taking strychnine or arsenic with their meals, as drinking whiskey, or gin, or even wine. There are thousands of families who oppose the use of any form of stimulants, even as medicine ; and there are those who cheerfully go down to death rather than defile their lips with the accursed thing.

In the crises of fevers, and in serious chronic diseases, physicians sometimes find great difficulty in inducing patients to take wine, or beer, or brandy even, when they are told that life is at stake, and stimulants offer the only hope. There are some physicians, even though their number is very limited, who refuse, absolutely, to prescribe any form of alcoholic stimulant, in any emergency.

THE BETTER CLASSES IN AMERICA MORE TEMPERATE THAN ANY OTHER CIVILIZED PEOPLE.

I speak of this habit of abstinence on the part of Americans, partially because it is so very recent, and because it is peculiar to this country. No such custom can be found in Europe, or in Asia. The better classes in the United States are the most temperate people that can be found in any civilized country.

They are the only Christian people who do not regularly use some **form** of stimulant **with their meals,** or at their **social gatherings.**

The reasons for this abstinence on the part of our substantial citizens are : *first,* the great agitation on the temperance question ; *secondly,* the nature of our climate, which will not allow free indulgence in alcoholic liquors ; and, *thirdly,* the fact that we have very little pure and cheap wine, and are obliged to use strong and usually adulterated liquors, that cannot be taken even in small quantities without injury.

Among the exceedingly rich and luxurious classes in our cities, total abstinence is more rare ; yet, even among these, instances of very great moderation are not unfrequent ; and even among the most extravagant and unprincipled of the wealthy classes drunkenness is looked upon with abhorrence, and total abstinence is rarely, if ever, ridiculed.

When, on the other hand, we come to the lower orders, and especially the ignorant and wretched Irish emigrants ; the poor whites of the South ; and the abjectly poor and despised of all nations, we find that intemperance is as frightful as in England and Ireland, among the same classes ; and that, as in England, it has *increased* of late years ; that the temperance reformation, and the general advance of the age, have had little or no influence upon them. They are on a plane so far below the ruling orders, that they cannot well be reached by moral influence, except in isolated cases.

According to the Police Reports of **New York** for 1868, of 78,451 arrests, 33,976, or nearly half, were put down to the cause of intoxication, or intoxication and disorderly conduct, and probably **one half of the other** arrests would never have happened if the victims had been sober.

The reports of the other large cities tell about the same story.

TEMPERANCE SOCIETIES AND TEMPERANCE LEGISLATION IN THE UNITED STATES.

Over eight centuries ago Mohammed prohibited wine to his disciples, and over four centuries ago temperance organizations of some kind were known to exist in Europe; but it was not until the beginning of the present century that the agitation on the subject assumed any very definite shape. One of the first to write on this subject was the very able physician, Dr. Benjamin Rush of Philadelphia.

The first temperance society in this country was formed at Saratoga, in 1808. It was local in its character and numbered 43 members, mostly farmers, who pledged themselves to abstinence from *distilled* liquors, and ordained that no member should be intoxicated under penalty of fifty cents.

In 1826 a Temperance Union was formed at Boston, but, like its predecessors, it allowed the use of wines, cider and malt liquors. The idea of total abstinence had not been entertained. A resolution of total abstinence from all that may intoxicate was proposed at a meeting in Philadelphia in 1833, and was voted down. The early advocates of temperance not only favored the use of wine, cider and beer, but also urged the cultivation of the vine as a temperance measure.

The doctrine of total abstinence was not adopted until 1836 at a convention at Saratoga Springs.

In 1840 seven hard drinkers met at a tavern in Baltimore and formed the order of Washingtonians.

XCISE AND PROHIBITORY LAWS IN THE UNITED STATES.

The custom of excise was brought to America from England by our Puritan ancestors. As early as the middle of the seventeenth century, the sale of alcoholic liquors in certain States, was attended with various and somewhat amusing restrictions.

The town of East Hampton, Long Island, for example, ordained that only those should sell liquor who were deputed to do so by the town ; that they should not sell to youths at unseasonable hours, and only one eighth of a quart to a person.

In 1788, New York passed a law punishing drunkenness with a fine of three shillings for each offence. That law has since been repealed, and now drunkenness is not a crime unless it leads to disorder.

New York for a long time had in force the license law, based on the acts of the colonies. It provides that in every town there should be a Board of Commissioners to grant licenses ; that only tavern keepers should have the privilege ; that the applicants should have good moral character, and that they should not sell to servants, or paupers, or apprentices ; and that any violation of the law should be punished by fines and imprisonment.

In 1845, New York passed a law providing that the majority of the electors of each town should have the privilege of deciding whether licenses should be granted in their town.

This law was repealed in 1847 without a fair trial.

In numberless ways the license law was evaded and with various success was fought in the courts. After two centuries of trial, it has done but little, if anything, to check intemperance.

In 1834, Congress passed a law, providing that any person selling strong liquors or wines to the Indians in the Indian territory, or introducing them into the Indian territory, should be fined ; that no distillery should be erected in there ; that any Superintendent of Indian affairs might on suspicion have the power of search and seizure.

THE MAINE LAW.

In 1851 Maine passed a *prohibitory* law, called the " Maine Law." Its leading features were absolute prohibition of the sale of alcoholic liquors as a beverage, with power of search, seizure and forfeiture of the liquor illegally kept for sale.

In every town an agent was appointed to sell liquor for medicinal or manufacturing purposes.

This law has been variously imitated and modified in Massachusetts and other States ; but, after considerable experiments has been mostly abandoned. It would appear to have met with more success in Maine than in any other State.

At the present time Connecticut and other States are proposing other experiments to reduce the liquor traffic.

From a careful review of these facts, we learn that :

Alcoholic drinks of some kind, mild or strong, pure or adulterated, are **used** *in about every portion of the globe.*

Already we have seen that some form **of stimulants** or narcotics could be found in all countries.   **This examination** shows that when we restrict our inquiries to *alcoholic stimulants,* the same universality **is found.**

It would seem, also, that the **use** of alcoholic liquors, **as of** stimulants and narcotics **in** general, **has** increased with the advance of the race.

They are **used** *to the* **greatest** *extent, in the widest variety, among those* **nations that are** *now leading modern civilization.*

In old and decaying civilizations, as Greece, Syria, Egypt, Italy and Spain, they are used much less than in young and rising nations.   Already we have seen that this is true of stimulants and narcotics in general ; it is just as true of alcoholic liquors.

They everywhere cause more or less of intoxication.

There appears to be no country where alcoholic liquors **are** used, **even of the** mildest forms—as the **purest and** mildest wines—where **there** cannot be **found,** among the **lower classes, at** least, those who habitually or occasionally indulge to excess and become either lively and silly, or ugly and beastly drunk.

This applies to all recorded ages.   Even the Egyptian hieroglyphics show that the oldest of the civilizations was **not a stranger** to intoxication.

Greece and Rome were fully cognizant **of** the vice, continually **refer to** it in their literature, and recognized its existence in their laws.

They are more **used and abused** *by some races than by others.*—The **Anglo-Saxons** and **their** descendants use them most; the Latin races the least.   The inhabitants of Great Britain, United States, Canada, Russia, Sweden, Norway and Denmark are most excessive with stronger liquors; while the inhabitants of France, Spain, Italy, South America and Asia prefer milder liquors, and use them in far greater moderation.

The Anglo-Saxons are the greatest colonizers, and wherever they go they carry intemperance with them.

The potent influence of race in determining the tendency to the abuse of alcoholic stimulants is seen in the comparatively little effect that the drunken habits of foreign residents and sailors have on the people whom they visit.   English and American sailors, wherever they go, carry with them habits of beastly intoxication, for sailors are the most intemperate class in the world; but they contaminate much less than would be supposed the innocent people who have little or no desire for the exhilarating effects of liquors.

Contrast the terrible figures of Liverpool, New York and London with the experience of Berlin and Paris. Contrast the general impression of those who live in those countries with the general impression of those who live in Germany and France.

Contrast the general impression on this subject of those who live in this country, and the general impression of those who live in France and Germany.

Judging from all the sources of information at my command, I should say that there is ten times as much intemperance (that is, intoxication from the use of alcoholic liquors) in proportion to the population, in Great Britain and the United States, as there is in Germany and France.

I am prepared to modify this estimate by any facts
that may hereafter be brought to light.

VALUE OF THE TESTIMONY OF PASSING TRAVELERS AS COMPARED WITH THAT OF RESIDENTS.

The testimony of ordinary travelers in Europe is
worth so very little to all truth-loving minds, that it may
be excluded from consideration.

We go through the Boulevards, or the Champs
Elysées, or the Unter den Linden in Berlin, and see no
men nor women reeling through those brilliant streets,
and forthwith write home that there is no intemperance
on the continent of Europe.

But how often do we see a drunkard in the upper
part of Brooklyn, or Fifth Avenue, in the day time?

Is it not possible to go up and down those streets a
hundred times and not see a man in the gutter, or trying to get there? And yet every year twenty or thirty
thousand are arrested in New York in a condition of
intoxication, and twice that number ought to be.

I saw the worst of Berlin, and, under the protection
of the police, I saw the worst of Paris; and I saw
only one drunkard; but the observation of any one
individual, however carefully made, amounts to little. I saw, however, raw material enough in Paris, out of
which drunkards are made; and, during the late war,
it has been called into activity.

By conversing with many old residents who have no
theory to form in this matter, (for, in some of these
countries they have never heard of the temperance cause,
having but little felt its necessity,) I find that the ignorant classes do get together, and become more or less

drunk, in France, in Germany, in Italy and Switzerland; but immeasurably less than they do in England, Scotland, Ireland, or America.

The testimony of foreign ministers is of considerable value, since they have opportunity, by long residence in one locality, to ascertain the habits of the people in that locality, that no passing traveler can possibly enjoy. Works of travel, written by men who evidently have no theories to form, and who have dwelt for a considerable time among the people whose habits they attempt to describe, are of great value.

COMPARATIVE INTEMPERANCE OF INDIANS AND NEGROES.

The relation of intemperance to race is very markedly seen in the different habits of the negroes and North American Indians. The North American Indians are notorious for their gross intemperance whenever they can get hold of alcoholic liquors. The same is true of the Kamschatkians. The African tribes in their native wilds are, some of them, beastly enough in their habits, and on their comparatively mild palm wine often become drunk and quarrelsome ; but bad as they are, they appear to advantage when compared with the American aborigines. In this country the ignorant negroes are certainly less intemperate than the ignorant Irish.

There is more of the grosser type of intemperance in the Northern than in the Southern climates.

It will be observed that the countries where the evils of intoxication are most felt, lie mostly above the parallel of 30° North. The nations above this parallel vary much in their habits ; but they include Great Britain,

United States, Canada, Netherlands, Sweden, Norway, and Russia, where intoxication is most severe and most observed.

The nations below this parallel, in both hemispheres, and on both sides of the equator, are vastly inferior to the nations above it, in mental or muscular force. They are either the remnants of the old and decayed civilizations, or are advanced savages, or in a state half way between savagery and civilization ; but they are not grossly intemperate ; and where they do indulge to excess they are far less ugly, quarrelsome, and disposed to crime, than the Northerners.

In the South there is more of the jolly drunk ; in the North there is more of the ugly, or dead drunk.

It would seem that none of the Eastern nations are grossly intemperate in the use of alcoholic stimulants.

The question whether this difference is due mostly to differences of race or of climate is one of vast complication, and suggests theories of the origin of species, and the relation of climate to character, which we cannot here discuss.

We may accept the general theory, that the differences of race are the result of differences of climate and other external conditions ; and therefore the difference in susceptibility of different nations to alcoholic stimulants is due, primarily, to difference of environment, and secondarily, to difference of race.

The food, also, which is a result of climate, has much to do with these differences of susceptibility.

Those whose food is meagre cannot bear stimulants as well as those who are liberally fed.

We have seen that extremely cold or hot climates do not well tolerate alcoholic stimulants.

The temperate zones bear them the best ; and there the stronger liquors are most used, and most abused.

Most of the ugly and beastly intemperance of Eastern and Southern countries is found among the foreign population.

The example of these seems to have much less influence for evil on the native population than we might suppose.

Since morality of all kinds is at a low ebb among most of these Southern and Eastern nations, we may urge that they are not tempted to drink to excess ; that they do not enjoy the exciting effects of stimulants as do the Northerners. It is rather as food, or as an accompaniment, or substitute for food, that the Southern and Eastern nations use their wines ; and the same may be said of some of the Northern nations.

Alcoholic stimulants are not as well borne, and sooner cause intoxication in North America than in Europe.

In this view there is pretty general agreement among all who have given the subject attention. Those who emigrate from Great Britain to Canada find that they must reduce their quantity of stimulants, whether they remain a short or a long time. That this is true of emigrants from different parts of Europe to the United States has long been known. In every part of the country this susceptibility is observed, from Maine to the entire Pacific coast.

It is least marked in the South, where more alcohol can be used than in the North, with less immediate effects. The native population in the latitude of New Orleans and Charleston use liquors more freely than the

native population in the latitude of New York and
Boston.

They drink oftener, and drink more ; the custom of
treating is more universal and tyrannical ; for the first
thing on being introduced to a stranger is to ask him
to take a drink.

It takes more liquor to make a man drunk in the
South than in the North ; and yet the Southerner bears
less than the inhabitant of northern Europe, certainly
less than the Scotchman. *Vice versa,* those who go from
the United States and Canada to Europe find that they
cannot, on short residence, use alcohol as freely as the
European with whom they associate ; on long residence
they acquire the same habits, which they must leave off
on returning to America.

The most obvious cause of this difference of susceptibility of Europeans and Americans is the greater *dryness of the air* of America.

THE PECULIAR DRYNESS OF THE AIR IN AMERICA.

The dryness of the air in America is indicated in
various ways.

1. The hair gets dry and harsh very easily, and large
quantities of oil and pomade are needed to keep it
smooth. This varies with different individuals, and
with the season. It is most observed during the
seasons when dry wind prevails, as in March and
November. For this reason, barbers are in greater
demand in America than in Europe.

2. Clothing that has been washed and hung out on
the line, dries quicker than in Europe. The German
women are astonished at this difference when they first

come to this country ; and that not more than half the time is required. On account of this ease of drying clothes it has become the custom to wash more frequently in America than in Europe.

3. There is less moisture on the windows. This difference is noticed especially in winter, when the frost flowers are less conspicuous.

4. Bread dries much more rapidly than in Europe. The custom of making bread every day, or every other day, and of depending on hot biscuit for each meal, is explained by this fact. The Europeans use more stale, and less warm bread, than the Americans.

5. Articles of various kinds mould less rapidly than in Europe. Hence the habit of using cellars for the storage of provisions. Prof. Desor, who is a naturalist of Neufchatel, states that, in Europe, collections of natural history are protected from dampness by lime and other absorbents. In America such precaution is not necessary.

6. Some of the different customs that prevail in this country are to be explained by the greater dryness of our air.

Paint dries more quickly, and the second coat can be sooner applied.

Cabinet makers, and makers of musical instruments, must be more careful in the selection of their wood.

Inlaid floors are so difficult to keep without cracking that they are quite rarely seen ; in Europe they are very frequent.

The superiority of the American pianos is supposed to be due partly to the same cause.

The plaster of newly built houses dries quicker than

in Europe, and the houses can be occupied sooner without risk of rheumatism.

Tanners observe that the skins of the tan yards dry more quickly here ; and therefore they can accomplish more in a given time.

6. The different appearance of the people. The English are stouter, heavier, have more fat on them, than their American descendants. The hair of the American is straighter, and his skin dryer, than that of the average European. The American neck is long and thin ; that of the European is thick and stout. This difference is observed most markedly in the female sex ; but it applies to both. It has been said that few Europeans grow fat in the United States, while many Americans increase very much in weight during a residence abroad.

Palfrey says that the curly hair, moist skin, and sanguine temperament so common in England, have in this country given way to straight hair, dry skin and the bilious or nervous temperament.

The skin of the American woman is softer, smoother, and thinner than that of the European : she is less gross ; her features are more delicately chiseled ; hence her greater beauty. There is here less development both of the muscular and of the glandular system. These differences have long been recognized. Those peculiarities of the American that distinguish him from his European brother—his great irritability, uncontrollable haste in business transactions, at meals, and in the streets, and his general nervousness—are to be explained mainly by the peculiarity of climate.

The enormous frequency of certain nervous diseases, such as nervous dyspepsia and the like, allows the same explanation; for our food, habits of eating and working, are largely the result of climate.

By these differences of climate, also, I explain the fact of observation, that literary, professional, and scientific men cannot work as many hours here as in Europe without breaking down ; that they work with less calmness and repose ; with a feverish eagerness of thought to which the European is a stranger.

Prof. Desor, to whom I am indebted for many of these facts and suggestions, states that the annual rainfall of the United States is equal to or greater than that of Europe ; and the number of rainy days is also not less than in Europe, with the exception, perhaps, of the British Islands and Norway.

The conclusion is, therefore, that during clear weather the air is less charged with moisture than in Europe ; and for this reason, that the prevailing winds blow over the land, while in Europe the prevailing winds blow over the sea, and bring moisture with them.

This explanation would appear to be applicable to pretty nearly the whole continent, though there are local differences of a decided character. It is stated, for example, that those who live on the border of the great lakes are particularly robust ; and by this view we may explain the fact that Canadians are less nervous than the residents of the United States, and that the residents of Maine, one sixth of whose surface is water, are so very robust.

RELATION OF ATMOSPHERIC MOISTURE TO ATMOSPHERIC ELECTRICITY.

The simple statement of the fact of the greater dryness of the air is hardly a sufficient explanation. It is at most **a begging of** the question. **Why does dry air** induce **nervousness?** An explanation that **I** have frequently **suggested,** and one that **has not** a little plausibility, **is, that in dry** air the **amount of** positive atmospheric **electricity is much less than** in air loaded with **moisture.**

I have here no space to speculate on this hypothesis, **(for it is nothing more,) and can** only make **the** statement that **the observations on** atmospheric electricity **that have been made in this country** and in Europe, so **far as they go, seem to confirm this theory.**

There is more positive electricity in the air in the forenoon than in the afternoon, **more** in the early evening **than just after midnight, more in** a fog than on a clear **day, more in a snow storm than in** a rain storm, more **just after and during a** thunder storm than just before it.*

Nervous and susceptible people are sometimes thrown into **convulsions and** attacks of vomiting just before a thunder **storm, and very many** suffer at such times from headache and neuralgic **pains.**

Intemperance varies with education and external conditions, and, in these days, works its greatest evils among the abjectly poor and ignorant classes.

A fact to which **I long ago** called attention, **but** which seems **not to be** appreciated, is, that most of **the** intemperance **of our day in** America and in Great Britain is **found** among the excessively **poor and** excessively ignorant classes, and it is an *effect* of this ignorance even more than it is a cause.

Ignorance is the fountain, intemperance is the stream ; the fountain flows back into the stream and makes it the more defiled ; but dry up the fountain and we dry up the stream.

Of about 20,000 arrests in Brooklyn in 1868, nearly 10,000 were born in Ireland, and no more need be said concerning their education. Of 78,000 arrests in the Metropolitian districts in New York in 1868, over 32,000 were born in Ireland, and 1,000 were negroes.

All nations that are even one degree removed beyond the state of savagery are divided into classes. Caste is universal. Everywhere there are Pariahs and Helots, and it is among these ignorant poor, and oftentimes oppressed and sickly peoples, that intemperance works its greatest ravages.

Of 63,000 arrests in London, only 17,000 could read and write well, and only 61, that is, one in a thousand, had superior instruction ; 2,000 could neither read nor write, and 84 could read only, or read and write imperfectly.

Of the 26,202 persons arrested in Liverpool in 1868, 222, or about one in a hundred, could read and write, and nearly eleven thousand or about half, could neither read nor write. Of those arrested for drunkenness and drunken disorder, 40 per cent could neither read nor write. Need I bring more figures than these ?

According to the Liverpool Report, drunkenness and ignorance would appear to be on the increase.

A certain proportion of those arrested drunk can read sometimes imperfectly ; but what does that amount to in the way of education ? What is it to be

able to spell **out** the printed page or scrawl your
name? **It is better** than nothing, and perhaps not
very much better ; for those **who** know so much on l**y,
very often read** only the lowest **literature, that** poisons
rather than purifies.

RELATION OF EDUCATION TO CRIME.

I do not forget that Herbert Spencer has argued very
forcibly and **very suggestively against the** theory that
gross **crime and education are incompatible.** This
able and original thinker arrays a certain number of
statistics, mostly of a local character, which go **to** show
that many of those arrested for crime are more or less
educated, and that among some of the classes thus educated crimes of a certain **kind are** relatively more frequent
than among **certain classes** below them **in intelligence.**
All this **I concede, and** still further, **I agree with** him
substantially **in** the view that ignorance **and crime not**
only stand to each other in the relation **of** daughter **to**
mother, but **are** frequently sisters, **having** a common
origin from a depressed social **state.**

The fallacy in Mr. Spencer's argument is that **he**
fails to make **the important** distinction between real
education, **that draws out** the powers and endues **one**
with higher aspirations and raises one into that plane
of social life where **the** incitements to virtue **are** far
weightier than the incitements to crime, **and** the simple ability to read and write, which is oftentimes worse
than no education at all. Success **in** some form of
crime requires a certain degree of intelligence, and the
ability to read and write is **an** accomplishment that is
pretty sure to be respected in an assembly of pickpock

ets, or professional burglars, or forgers, or counterfeiters. It is so much capital in trade, and is valued accordingly. Therefore we need not be surprised to find that the majority of the inmates of our prisons have attained to that degree of culture where they can scrawl their own names and read the Police Gazette.

Of 2,422 criminals confined in the Philadelphia County prison at different times and for various grave offences between 1855 and 1863, 1,737, or over 70 per cent could either read, or write, or both ; and on comparing these reports with those of other portions of the country, I find no evidence that this culture was in any way superior to the average of criminals.

We should not forget that *many of the crimes are trades.* Like all other trades, they are learned only by early initiation, and by long years of practice. Those who follow them are organized in societies and bands like honest artisans, and like them, also, are subordinate to their own laws and regulations. A man who attempts picking pockets, robbing banks, or stealing railroads, or bribing judges, on his own hook, without preliminary training, will be as sure to fail as though he attempted to extemporize himself into a carpenter or a jeweler, a lawyer or a physician.

There are in all civilized countries, but especially in Great Britain and in the United States, a large class who are not smart enough to be criminals of the first rank; if they wish to disturb the peace of society it must be by second-rate methods. There are a few crimes that are just as well committed by those who are unable to read and write— and who, perhaps, have not the ability to acquire such knowledge—as by those whose

endowments are more liberal.   Intemperance is one of
the chief of these crimes.

INTEMPERANCE A MATTER OF MORAL CHEMISTRY.

Bring together fat and alkali, under certain conditions, and you make soap.   Bring together ignorance,
extreme poverty, and alcohol, and in certain organizations, among certain races, and in certain climates,
and you make a drunkard ; and the law is no more in-
evitable in the one case than in the other.

*Intemperance is a matter of sex ; women are not as
intemperate as men.*

Abandoned women—thieves and vagabonds—are like
the males of the same class, intemperate.   Half of the
32,000 arrested for intoxication in New York, were females.

In Liverpool, in one year, more females than males
were arrested for intoxication.   In London there is not
much difference.

Feminine intemperance among the intellectual and
refined classes, on the other hand, is one of the rarest
of offences in every civilized country ; and the newspaper reports to the contrary, are nothing more than
psychological studies.   It would appear that woman
uses less tobacco, and less of many of the grosser
forms of stimulants than man.   Her finer organization
repels these things ; not so much because her moral
force is greater, but because her temptation is less.

This remark appears to apply to nearly all countries ;*
but with especial force to high civilizations.   Among

---

* Du Chaillu states that among the Apingi tribe of Africa, who get drunk on
palm wine, the women " are much more temperate " than the men.

the Greeks and Romans the women used less wine than the men ; and in Rome wine-drinking by women was regarded as disreputable.

The comparative exemption of woman from intoxication from alcohol, among all except the lowest orders, is all the more remarkable from the fact that o some other stimulants and narcotics she is even more addicted than man. Tea and coffee, especially the former, woman worships with far the stronger devotion.

Intemperance varies much less with religion and government than might be supposed.

The English Protestant and the Irish Catholic forget their differences of creed, and roll together in the same gutter. Austria is nominally Catholic, and Prussia is nominally Protestant, and both are, as nations, temperate. France, half Catholic and half rationalistic, is more temperate than Protestant Scotland and England. America has all religions, and all the intemperance it can well bear. A good religious creed is certainly, on the whole, conducive to morality ; but the class of people who keep the ranks of drunkards full, usually care very little for religion, except for the superstitious element in it.

That Protestant Christianity has been, directly and indirectly, a factor in bringing about among the better classes the great temperance reform, there is no question.

And yet it must be confessed that, with the exception of the Irish and Russian peasantry, the most intemperate people are to be found among the nominally Protestant.

One of the founders of the temperance reformation

in Great Britain has observed that his attention was first called to the subject when traveling in France, where he was struck with the fact that the French peasantry—Catholics in their faith, or else free-thinkers—were yet far less intemperate than the Scottish peasantry, who were rigid Presbyterians.

In the light of our present knowledge of the subject, we see that race and climate, more than religion, were the causes of the greater intemperance of the Scotch.

THE RELATION OF MOHAMMEDANISM TO INTEMPERANCE.

Mohammedanism appears, thus far, to have done more to prevent intemperance than any other religion. The Prophet forbade the use of wine, (although he himself indulged in it,) and the injunction is, as a rule, observed. Whatever Mohammedans do in the way of drinking alcoholic liquor is done furtively and usually in moderation. In the north of Africa, where the date-tree wine (palm wine or "lagmi") is in use the Mohammedans do not hesitate to indulge in it; for by a quibble, which is both amusing and untrue, they contend that the lagmi is not wine, and therefore does not come within the prohibition.

Next to Mohammedanism Brahminism appears to have been more or less successful in imposing the doctrine of total abstinence. It should be considered, however, that both of these religions belong to a very superstitious and degraded people, who are more influenced by the teaching of their leaders than Christians of any class, and a people too, who neither by race nor by climate would be predestined to intemperance. Protestant Christianity is the religion, on the other hand, of a liberty-loving and alcohol-loving people, who have a strong passion for independence, and for sensual indulgence in its most active and violent forms.

THE HEBREWS A TEMPERATE PEOPLE.

The Hebrews also are a very temperate people. Amid all the persecutions and vicissitudes from which they have suffered, they have never, as a people, in any climate, been driven to intemperance. Many other races, similarly distressed, would have taken refuge in the wine cup. Not only are they a deeply religious people, but they belong to a temperate race, and they are very rarely illiterate.

RELATION OF FORMS OF GOVERNMENT TO INTEMPERANCE.

He who should attempt to trace a relation between the forms of governments and intemperance would find that he had encountered a difficult task. Republican America, liberal England, despotic Russia are all intemperate, while Prussia and Austria, with their absolutism, and France with her anarchy, and alternations of license and tyranny, are comparatively temperate.

The only generalization that can be drawn is that the freest nations are the most intemperate, for Switzerland is less temperate than the despotisms by which she is surrounded, and no nations are lower in the scale of drunkenness than Great Britain and the United States.

On the other hand, the despotisms of Asia and Africa

are seen to discourage the immoderate use of alcoholic liquors more effectually than liberalism or republicism.

Intemperance of the gross form in the countries where it exists causes a very large percentage of the gross crimes of those countries.

In some countries a pretty large proportion of the arrests for assault and battery, for violence of various kinds, for disorder, and for positive murder are in some way associated with intemperance.

It would be unreasonable to infer that the absence of stimulants would proportionally reduce the gross crimes of those countries, for it should be borne in mind that the tendencies to crime vary with race and climate, and other conditions, like the tendencies to intoxication, and the world over, the most intemperate people are the most criminal, whether they are drunk or sober. Their crime is the effect of their temperaments, and intemperance is one of those effects; and one which both results in and aggravates all the rest.

Some of the most wretched people of the world are yet not criminally disposed. There is more of assault and battery, and of murder, and of ugly, bloody quarreling among the English, even when they are sober, than among some of the half civilized people.

Life and property seem to be safer in many of the islands of the ocean, among the ruins of the old civilization, than in London or New York.

The inference is that we are to beware of the tendency to attribute all of our crimes to intemperance, and all the peace and quietness of these other and inferior people to their temperance.

CORRELATION OF CRIMES.

The qualities of the mind that dispose to crime seem to stand toward each other in a kind of correlation.

The study of the mythology of a people greatly aids in determining the national character. As the gods of Greece partook of the refinement and poetry of the Grecian character, so the Scandinavian mythology, with mighty Thor and his huge hammer and nameless varieties of gods, with their riotings, and feastings, and bloody battles, represent the energy, the passion, the courage and executive force of the Northmen.

The English, the Germans, and the Americans are, of all people, the most energetic. Associated with this courage and vigor, is a powerful development of some of the lower passions. They are fond of the pleasures of the table ; they are great eaters and drinkers. They are active colonizers, going everywhere, and carry with them their habits of sensual indulgence, their fiercer passions, their coarseness, and their vulgarity.

Less licentious, less artful, less intriguing than the Frenchman, the Italian and the Spaniard, they are far more addicted to coarse and brutal crimes. Crimes of violence, as murder and murderous assaults, would appear to be more frequent in Great Britain than in the Southern countries, as among the negroes, or the South Americans. Those who study the statistics of crime in its relation to intemperance should bear in mind that the most intemperate people are those who are most disposed to crime even when sober.

Coarse crime and drunkenness are twigs growing on the same stem.

In all society, and especially in civilization, men live

in constant restraint; the lower and degrading passions are kept under, and we see the man, not as he is, but as he thinks he ought to be. When a man is drunk he loses hold of the reins of his mind, and the noisy rabble of passions run riot as they please, and we see the man as he is by organization. Hence the reason of the maxim " *In vino* **veritas.**" The energetic, **courageous** northern people when drunk are savage, ugly, brutal, and inclined to swear, stab and murder; the timid, indolent and easy southern people, when drunk, are jolly, lively, gay and garrulous. **Herein we** find **the** explanation of the paradox that intemperance is **most frequent and most destructive** among the most powerful, **the** most civilized and the most cultivated nations.

The **cultivated Englishman or American of to-day is a very different being from the** Anglo-Saxon, **from** which he sprung.

It is possible to **see** something of **the old Anglo-Saxon** element in the cheap and popular eating saloons of London and New York, and on **any** English steamers, where provisions are served **in** a coarseness and abundance that would have rejoiced **the heroes** of Valhalla.

*The effect of pure native wines, in countries where they are produced, **appears to** diminish intemperance among the classes who **use them** pretty exclusively, but **only** within certain limits.*

There are very few countries where wine is produced **that are** not also able to obtain nearly **all** other forms of alcoholic stimulants; and if by race or climate the inhabitants are disposed to **intemperance,** the wines will not entirely prevent them, even though they are perfectly pure, and absolutely cheap, and universally accessible.

It has been observed that when wines become scarce in countries where they have been cheap, the effect is to increase intemperance to a certain degree. In countries such as Switzerland, where some districts are liberally, and others scantily supplied with wines, the former are the more temperate.

In countries where race, and climate, and customs all urge to intemperance, abundance of pure wine has little influence to counteract that tendency. Pure wines will not satisfy, and they search for stronger liquors, and will get them and abuse them.

Strikingly this is illustrated in California, in France, and in some of the Rhine provinces.

What is true of wine is true also of beer and cider. In Great Britain beer is very cheap, and ordinary wine is not very dear ; but intemperance is frightful.

The intemperance of a people is not always in proportion to the quantity of alcoholic liquors that they use.

England uses more alcohol, in various forms, than Ireland ; but it has less drunkenness. During the last quarter of a century the use of alcoholic liquors has increased in Denmark ; but, at the same time, intemperance has diminished.

In France, the consumption of alcoholic liquors— not only of wines, but of the stronger varieties—is very great; but the French are by no means a grossly intemperate people.

The explanation of this paradox is that national intemperance does not result so much from widely diffused habits of drinking, as from great excess among a limited number. A certain portion of the people—

usually the abjectly poor and degraded—may be drunkards, while the better classes are teetotallers.

To this may be added the consideration that in some climates large quantities of alcoholic liquors can be taken habitually without producing intoxication.

Intemperance increases in times of war and great excitement.

Amid arms, morals as well as the laws are silent. Profanity, robbery, and licentiousness, and with them intemperance, usually follow in the track of great wars, and continue to infest a people years after the declaration of peace. In this, as in many other features, the evil that war does lives after it. Our recent civil conflict opened upon the land a flood of intemperance that has not yet subsided. During the past ten years the vice has certainly been more wide-spread among the wealthy classes, if not among the middle classes, than during the ten years just preceding.

According to all accounts, the late Franco-Prussian war has been attended by great increase of intoxication, even in Paris, one of the most temperate of cities.

In general, it may be said that crises that foster coarse vices will foster intemperance.

The formal and systematic agitation of the subject of intemperance is entirely of modern birth, and is now confined to a few countries in which the evils of intemperance are most obvious.

Prior to the last half century temperance societies had been but little thought of ; and now they are found chiefly in Great Britain and United States. There is, as we have seen, one society of the kind in Netherlands, and one in Switzerland ; but these are not established on total abstinence foundations, but are devoted mainly to the abolition of the custom of using strong liquors, such as rum and gin, which in those countries do not a little injury among certain classes, though far less than in America and Great Britain.

In Germany, in France, in Spain, and Italy, the subject has not been formally agitated, and has never been regarded as a great social question.

This may be explained by the consideration that moral reforms of all kinds are more active in the United States and Great Britain, but mainly by the fact that in these countries the need of a temperance reform has been most severely felt.

Greece, and Spain, and Italy have needed societies for the suppression of brigandage, and licentiousness, and the encouragement of common morality ; but they have not greatly needed the temperance pledge.

The success of the Temperance Reformation, though very remarkable, has thus far been mostly confined to the intelligent classes, and among the abjectly poor and ignorant, intemperance has at the same time increased, even in those countries where the agitation on the subject and the results of agitation have been most decided.

That the great middle and upper classes of Great Britain, and the leading classes of the United States are more temperate now than they were half a century ago, there is no question. It is conceded by the temperance reformers, and by philanthropists of all kinds who have had opportunity for comparative observation.

On the other hand, it is just as well demonstrated by statistics and by observation that among the wretchedly poor and ignorant classes of these countries, intemperance has at the same time increased.

The enemy has changed his base, but he is not conquered.

It may be doubted whether there was ever a time in the history of the world when there was so much gross intemperance as at present.

The following statistics, showing the proportion of those apprehended for "drunkenness and disorder," to the population in certain prominent towns of England, are taken from Graham's Temperance Almanac for 1869 and 1870.

Towns.				Towns.			
London,	1	in	169	Newcastle,	1	"	62
Liverpool,	1	"	30	Rochdale,	1	"	52
Manchester,	1	"	38	Exeter,	1	"	803
Birmingham,	1	"	128	Oxford,	1	"	2,469
Leeds,	1	"	161	Warrington,	1	"	31
Sheffield,	1	"	181	Norwich,	1	"	727
Bradford,	1	"	477				

It will be observed that the manufacturing towns present the worst exhibit, while Exeter, Oxford and Norwich are comparatively temperate.

Statistics also show an increase of the number of arrests for drunkenness and disorder from 1867 to 1868, in many of principal towns of England. It will be seen that the increase is much too great to be accounted for by the increase of population, and that it applies to all the towns, large and small.

Number of arrests in 1867.		In 1868.
London	16,608	18,872
Liverpool	11,938	14,451
Birmingham	1,330	2,310
Leeds	1,340	1,369
Sheffield	841	1,022
Bradford	191	285
Norwich	93	103
Bolton	729	1,217
York	208	289
Rochdale	629	743
Exeter	42	431

Prohibitory legislation, under any government, has thus far been no more successful against alcoholic liquors than against other stimulants and narcotics.

We have already seen that tobacco, coffee, coca and even tea, have in various ways met with fierce and violent opposition ; that not only moral, social and ecclesiastical, but, in a number of instances, governmental influences, have been arrayed against them. We have seen that despotisms and republics, differing in nearly all other features, have agreed in this, that the use of these agents should be repressed by law. We have seen that almost always and almost everywhere they have signally failed. Despots, who held the lives and even the thoughts of their subjects at their bidding, who at will controlled all their outgoings and incomings, their dress, their thoughts and their prayers, could not restrain them from the habitual use of stimulants and narcotics.

Nowhere is government more absolute, nowhere is it more reverenced than in China ; but even there, in the warfare against opium, the people have gained the victory. Prohibitory legislation against alcohol is of later date than prohibitory legislation against several other substances,* but until quite recently, tobacco, coffee and tea were regarded as far more pernicious. While kings and popes were fulminating against tobacco, alcoholic liquors were then, as now, the habitual drink of the people, but were not regarded with apprehension.

Temporary, local, circumscribed successes have certainly been gained by legislation ; but little or no permanent relief from the great evil of intemperance.

The stream has been dammed at certain points, and for a time men thought it had been stopped, but the overflowing of the surrounding country, and the subsequent breaking away of the dam with great destruction of life and property, has brought us to the inquiry whether it would not be better to attack the source of the stream.

How shall we prevent and cure intemperance?

The doctors love to say, and quite right we are in so saying, that the best test of medical skill is in making a diagnosis, in finding out what the matter is with our patient. After that is done, very likely any old granny can give the medicine.

---

* Pittacus made a law that a double punishment should be accorded to him who committed a crime when drunk—one for the crime and the other for the drunkenness.

Solon made drunkenness in a magistrate a capital crime.

The Roman law was lenient to men who committed crime while intoxicated, but punished drunken women with death. Carthage forbade even the cultivation of the vine. Charlemagne ordered that drunkards should be punished for the first offence by private—for the second by public scourging, and if that failed, by death.

In the time of William and Mary, drunkenness was punished by fine and imprisonment.

The medicine may not be a specific, may be very falli-
ble, and yet be just the best we have, and all may know
of it, and how to use it.

Medical science has very few specifics.   Our best and
most used remedies—opium, hydrate of chloral, quinine
and electricity—are none of them certain to cure even
any one disease.

NO SPECIFIC FOR INTEMPERANCE.

There is no specific for intemperance. It follows
from the *diagnosis* I have given of the nature, causes
and habitat of intemperance, that it must be treated con-
stitutionally, and locally, and by a variety of measures—
all of which may do some good, but none of which
can do all that a philanthropist could wish. Would
that some moral quinine could be found that should
break up these attacks of drunkenness! Would that
by some process of vaccination we could save our
youth, so that they should never have the disease, or
only a mild form.

I have said that intemperance is partly a vice and
partly a disease ; in either case the drunkard is to be
pitied more than blamed.

THE GREAT THING IS PREVENTION.

Like all other vices and diseases, intemperance
when chronic is very hard to cure, and our main efforts
must be directed to its prevention.

Lay the axe at the root of the tree. Throw all the
energies of society into the attempt to make it impossible for any one to be born and grow up in this country

without knowing how to read good books.   Reduce ignorance to a minimum and we shall do much toward re-
ducing intemperance to a minimum.   Even the intemperance of refined and cultivated society, results
mainly from ignorance, and the recklessness, and
slavery to passion that comes from ignorance; for educated people in this country have known but little or nothing of the history and effects of stimulants
and narcotics, and withal most of them in these classes
who became drunkards, form their habits early, before
twenty-one, while as yet they knew little of science or of
themselves.

It cannot be too often repeated, line upon line and
precept upon precept, and here a little, and there a
little, that gross vice and ignorance go together.

Intemperance, in short, must be treated like other
great social vices, like the vice of licentiousness, like
the vice of lying, which is more wide spread than
either.

By every possible means raise the tone of society and
coarser vices will disappear.

This is the constitutional treatment; at the same time
let local measures be applied, for between them there
is no interference.   An advantage of this constitutional
treatment is, that while it diminishes intemperance, it
at the same time causes also other associated vices to
disappear.

LOCAL TREATMENT.

By popular gatherings, by the circulation of special knowledge on this subject through the pulpit
and the press; by science, by art, by organization, by

law, so far as law proves to be of service in checking
the coarser manifestations of the evils, by all these means,
not singly, not alternately, not interruptedly, but uni-
tedly and simultaneously we should attack this great evil.

In the past we have had too much local, too little of
general treatment ; too much dependence on law, on
" coat-tail eloquence," on pledges. The process I recom-
mend is a slow one, this gradual lifting up of the people, but the mills of God always grind slowly. There is
no short-cut to national virtue. The laws of reform
are as fixed as those that control the march of the stars.
There is no patent medicine which an intemperate nation can take and be healed.

Even were a prohibitory law enforced, one might
honestly doubt whether it would do very much to dry
up the fountain.

Against the social evil, law has failed. Statistics
have shown that in France, and other countries where
it has been tried, it has not diminished either the vice
or the diseases that result from the vice ; although
in Great Britain local successes have been gained.

With the march of progress, the social evil has, on the
whole, wonderfully diminished ; it is incomparably less
than it was six centuries ago.

The genius who invents a law which without oppress-
ing the majority shall restrain those who are disposed to intemperance, so as to give the power of education better chances to work, will have as many stat-
ues built to his memory as he cares for ; but that
genius must have a thoroughly original mind.   He
must not copy after anything that has yet been attempted.

The best temperance law would be an act compelling all children between certain ages to attend some good school.

If we make it impossible for children to grow up ignorant, we make it improbable that they will grow up intemperate.

Every time we open a school-house we **close a dram-shop.** The schoolmaster is the rumsellers' worst enemy.

In the management of our schools, in the methods of instruction and in the methods of discipline there is vast need of improvement.

The time which is now spent in teaching children the names of cruel and insane, or idiotic kings, and the dates of their reigns, and the number of men they killed, should be devoted to teaching them the laws of their bodies, the principles and the facts of hygiene, and what is really useful to learn in history and general science. Much of history, as usually taught to children, is too much like a bound **Police Gazette,** a condensed Jack Sheppard.

If children could be taught simply to read a decent newspaper, to know the common laws of life, the vice and crime and misery of the land would be very greatly diminished, and intemperance would probably be reduced a large percentage if the young could be taught the history, the nature and the effects of stimulants and narcotics in such a way that they could appreciate and believe what was taught them.

The need of improvement in the *methods* of instruction is equally imperative. There should be less of the abstract and more of the concrete, less of the disagreeable and more of the agreeable, so that to the average of youthful minds the school may bring pleasurable rather than painful associations.

These are generalities, but to enter into details of the methods of education would be inconsistent with the scope of this treatise.

It must needs be a long time before the system of primary instruction can be made what it should be. First of all it is necessary to remodel the system of higher education, a process which has just commenced; but which is now going on with reasonable rapidity and which is sure to be accomplished.   We must teach the teachers of our teachers, and the reformation of our primary schools will surely follow.

A law making education compulsory, if passed before the whole land becomes steeped in ignorance, can be well enforced in most sections, and the beneficent effects of such a law is shown in Prussia, where knowledge is not only better concentrated but better diffused than in any other country.

ENORMOUS AND INCREASING ILLITERACY IN THE UNITED STATES.

There are in this country *four and a half millions of adults and youths over ten years of age* who cannot read or write, and probably ten millions more whose reading is only of the lowest order.   This illiteracy has been increasing for the past thirty years, not only among the foreigners, but among the native population even in New England.

Those who argue with Herbert Spencer that the duty of the State is simply to protect, that it ·ought not to

educate, and, strictly speaking, cannot educate at all if
it tries, may be met, *first*, by the consideration that of
all enemies of the State, ignorance is the greatest, since
it gives birth to criminals, to traitors and to paupers,
and, therefore, the State should be protected from it at
every point, and that intelligence is a better national
defence than standing armies or iron-clad navies; and
secondly, by the fact already mentioned, that States *have*
succeeded in educating the people.

We can make the men abstain, either by taking the
rum from them, or taking them from the rum. The
inmates of our prisons are teetotallers; but is that the
character which makes a great nation? We want a
virtue that can march through files of dram shops, and
be untempted and unharmed.

EFFECT OF STOPPING THE GROG RATION IN THE NAVY.

During our late war, by the influence of Admiral
Foote, the grog ration was taken from the navy. I
know not how bad things were before, but imagination
could hardly paint them worse than they were after the
enforcement of that law. The carrying out of the law
on the sailors was a success, for the grog ration was
simply stopped; but as assistant surgeon in the navy
at that time, I had all the chance I wished, to study the
relation between law and virtue. At sea, sailors were
sober for the same reason that the inmates of Sing Sing
are sober; but when the vessel went into port, with a
unanimity that was both amusing and amazing, they
became drunk, and, so far as they dared, kept so.

We must avail ourselves of the weaknesses of human

nature, and make intemperance unfashionable. We
have done that already, both in this country and in
England.

Emerson says that "moral qualities rule the world,
but at short distances the senses are despotic;" and
again he says, that he would rather sit at table with a
man who did not obey the laws than with one who is
repulsive in his personal habits.

We may say that to one man who does right from obedience to God, ten do right from obedience to fashion
or their senses.

The habits of tobacco chewing and profane swearing
are steadily disappearing from the Eastern and Northern
portions of this country, where the best culture lies, not
because God, but because society—the ladies—forbid
them.

On the other hand, among the lower orders—among
the degraded—down into those depths where the light
and wealth of civilization do not enter—all these vices
are as frequent as ever they were ; and among these
classes intemperance, in spite of all our reforms, is increasing—certainly in England, as statistics show, and
probably, in this country also, among the corresponding classes.

We must also attack ignorance in the higher classes.
We must spread abroad sound, thorough, impartial
knowledge on this special subject.   There was a
Roman proverb, "when vice is rewarded it is a crime to
be virtuous."   So long as riches are honored, even when
gained by violence, so long will there be ignorance in
high as well as in low places.

INEBRIATE ASYLUMS.

The intemperance that is a disease, and even that which is a vice, is oftentimes best treated by some form of restraint ; and I am strongly in favor of inebriate asylums. They are not perfect institutions, but they are good ; and in the main, I believe, worthy of confidence, and in a certain proportion of cases they cure for ever. Let no drunkard utterly despair until he has tried them.

The cases of spontaneous permanent recovery from chronic intemperance, like the cases of recovery from consumption, are rare ; the great task is to treat the early stages.

High on our temperance banner raise the Roman motto, "principiis obsta," stop the beginnings?

It follows, from what I have said, that the introduction and popularization of native cheap wine would not do away with intemperance, although it might modify it. Turn the oceans into beer, let all our rivers run with wine, and so long as we have abject ignorance and abject poverty so long we shall have drunkards.

CHAPTER IV.

STIMULANTS AND NARCOTICS IN THEIR MORAL, SOCIAL, AND ECONOMIC RELATIONS, WITH PRACTICAL SUGGESTIONS CONCERNING THEIR USE.

WE come now to this broad question :

Have stimulants and narcotics, on the whole, been a benefit to the human race?

The scope of this question will be better perceived by varying somewhat its phraseology. Is the *average* result of the use of these agents on the human family beneficial? Are the undoubted and terrible evils that, in numerous individual cases, come from some of them in many countries—and which in some sections and orders of society are certainly increasing—more than counterbalanced by the pleasure and profit that is derived from cautious and temperate indulgence in them?

To this general question it would seem impossible to give any other than a decidedly affirmative answer.

The simplest, and shortest, and most certain solution of this problem would be found in a world constituted just like ours, with similar varieties of race, climate, similar in history, external conditions, food, and in every feature except in the use of stimulants and narcotics. A comparison thus made would afford us an answer that would be absolutely convincing, leaving no room for query. But the pathway to truth is not usually so clear or well paved ; it is narrow and crooked, cut up into various broadways and crossings, so that at every step we may go astray ; and oftentimes the best result of the hardest tramping and toiling through many devious by-paths, where we have often lost our way, is merely an approximation to the temple of truth. We come near enough to get a misty view of its exterior, and to trace its general outline, but are not permitted to enter the golden gates.

There are two or three general considerations that have not received the attention they seem to deserve, and which very much aid us in answering this question. These are—

1. *The world over, with some local exceptions, woman uses less of stimulants and narcotics than man ; and yet she is weaker than man.*

THE WOMEN OF THE UNITED STATES.

There is no civilized country where the women of the ruling class use less alcohol and tobacco than in the United States, and there is no country in the world where the women are so frail. They are relatively very frail in comparison with the male sex, the difference in the health of the sexes, in regard to health, being greater than in any other country.

The American woman is nowhere stronger, but usually weaker. She seems to have more sickness, certainly more of diseases, of debility and exhaustion.

There are some countries where the use of some one stimulant is almost as frequent among women as among men, and in these countries the relative weakness of women is less distinctly observed than in countries where the sterner sex mostly monopolize these substances.

The German women use beer and wine in great abundance, like the men, and are stronger than the women of any other civilized country.

The English women are incomparably stronger and more enduring than the American, and they use wine and beer to which the average American woman is almost a stranger.

In India and Siam smoking is the amusement of all classes and both sexes.

In Burmah all ranks, and classes, and both sexes, even to little infants of three years, smoke cigars.

In China, it is said that every female from eight years and upwards wears on her clothes a little pocket to hold her pipe and tobacco.

WOMEN IN BARBAROUS LANDS.

It is only in barbarous or semi-civilized lands that woman uses the same stimulants and narcotics as man, and it is in those lands that she bears children with least difficulty, has the least of nervous disease, and is most capable of severe toil. Woman has suffered more from civilization than man, and civilization has caused her to abstain from many of the stronger forms of stimulants and narcotics which man now uses more freely than before.

There is no civilized country where both sexes preserve the same habits in their use of stimulants and narcotics ; for even where tea and coffee, wine and beer, brandy and whiskey are used by both alike, smoking or chewing is a privilege which usually man seems to reserve for himself.

The use of tobacco is one of the "*rights*" that woman most cheerfully foregoes. The custom of smoking pipes, once quite common among the sturdy women of this country, is now almost obsolete.

If the women who thus abstain from all forms of stimulants and narcotics except tea, chocolate and coffee, were stronger and longer-lived than their husbands, while those who did not thus abstain were weaker and shorter-lived, we should have an argument very potent against the use of tobacco and alcoholic liquors.

We find the case to be directly the reverse, and it is not unreasonable to infer that these substances are not in their average result pernicious.

2. *The use of many forms of stimulants and narcotics has increased many fold with the advance of civilization, and especially in those countries that march at the head of civilization, and, at the same time, longevity has increased, and is nowhere greater than in Germany, Great Britain, and United States.*

To argue from this coincidence that the use of alcohol and tobacco, tea and coffee, has been the sole or principal cause of this increase of longevity would be a kind of reasoning the very reverse of scientific, for the reasons for this increase of longevity are manifold ; but we are surely not far wrong in inferring that the average effect of this increasing use of stimulants and narcotics has not, to say the least, very greatly shortened the average longevity.

3. *Those few in civilized lands who entirely abstain from all stimulants and narcotics seen to be no healthier than those who make more or less use of them.* The number of total abstainers is so limited that any extended comparison is difficult, and we are obliged to form our judgment from isolated cases. There are individuals who attain great age on positive food alone, or on very little of negative food, and their examples are frequently cited as arguments in favor of entire abstinence from all these things ; but we forget that far more instances of equally good health and great longevity are to be found among the habitual users of stimulants and narcotics.

4. *It is during the past half century that the doctrine of total abstinence has arisen, and has been adopted by a very large portion of the better classes of the United States, and during this same period there has been a more rapid increase and multiplicity of nervous diseases than in any other period of the world, and this increase has been mainly among the temperate classes.*

That there has been such an increase of nervous diseases, during the last quarter of a century at least, is shown both by statistics and by the general observation of physicians.

It would be extremely unscientific to conclude that this increase is due entirely or largely to total abstinence ; for the elements of climate, severe brain work and worry, are probably the leading causes of this increase : but this much may at least be claimed, that approximate or absolute total abstinence among a very large class of the American people has not kept them from becoming **the** most nervous people on the globe.

THE COMPARATIVE INFREQUENCY OF NERVOUS DISEASES AMONG THOSE WHO USE ALCOHOLIC LIQUORS TO EXCESS.

Inflammatory diseases of various kinds are **excited**, and aggravated, and made more fatal by free indulgence in alcohol, as is shown by general observation **and by** statistics ; **but nervous** diseases are certainly no **more** frequent, **probably less** frequent, among **hard** drinkers than among the total abstainers.   **In this view** both Magnus **Huss** and Marcet accord.

But without the statistical evidence of these authorities, **common** observation might easily enough settle the **question.**   We find in this country the most of paralysis, of neuralgia, **of hypochondria,** of hysteria, of chorea, of nervous dyspepsia, **of insomnia, of nervous** exhaustion, not among **the** intemperate foreigners, nor among the intemperate classes of any nationality, **but** among the upper and middle classes, **who are either** teetotalers or moderate and occasional drinkers.

Since alcohol goes to the brain, **and makes** its influ-ence chiefly felt **through the nervous system,** it would be natural **to suppose that** drunkards would die of nervous diseases ; **but such is** not the case.

Febrile and inflammatory diseases go hard with them, and prevent them from attaining a high average **longevity.**

That certain forms of stimulants **and** narcotics have done **more** evil than good to **certain** localities is unquestionable.   China, for example, would have been healthier if she had never known opium, and in this

country ardent spirits have probably wrought more of
mischief than of benefit to society.

THEY ARE ANTIDOTES TO EACH OTHER.

One fact of importance to be considered, is that the
various stimulants and narcotics, though in the main
similar in their effects, yet differ widely in special features.   They may be antidotes to each other, and
to a certain extent substitutes for each other.

As belladonna is an antidote for opium, although
both affect the nervous system in some respects similarly, so to a certain extent coffee is an antidote for
alcohol, and therefore the custom of taking a cup after
dinner is a wise one.

It is stated by Mr. Lecky, that by the introduction of
coffee into France in the 17th century, drunkenness,
which had been increasing to an alarming degree
among the French, was quite effectively checked. There
is no doubt that coffee has co-operated with Mohammedanism and climate and race, to keep some of the
nations of the East so remarkably temperate.

There is no doubt that tobacco and tea and coffee
and alcohol combined, have done very much to save
Europe and America from the opium-eating habits of
the Eastern nations.

The increase of opium-eating in this country and in
England since the agitation of the temperance question, has been very great.

In the United States, between 1840 and 1867, the
population increased in the proportion of 210:100,
and the importation of opium increased in the proportion of 650:100. In the opinion of those best quali-

fied to judge, four fifths of this is eaten. In 1869, 30,000,000 pounds of opium were imported. In some of the counties of England, opium-eating has become a very serious habit.

STIMULANTS AND NARCOTICS VERY EXPENSIVE.

The costliness of stimulants and narcotics as compared with ordinary food, does not seem to have attracted the attention it deserves, even from temperance reformers.

We have seen that in Ireland each family spends *fifty dollars annually* for alcoholic liquors ; that in Austria and Hungary, eight times as much is spent for the same substance as for iron.

According to recent statements, the people of the United Kingdom spend annually on wines, spirits and malt liquors, over $500,000,000, or over *sixteen dollars for each individual.* This is about eight times as much as they spend for cotton goods.

In this country, where liquors of all kinds are much more expensive than in Europe, Mr. E. Young has estimated that about $650,000,000 worth are annually sold over the counter, which is *over sixteen dollars to each inhabitant.*

When to this we add the expense for tea, coffee, tobacco and opium, we find that our negative food is a most important item in the history of a highly civilized nation.

Among the semi-civilized and barbarous nations of southern Europe, the stimulants and narcotics in common use are comparatively cheap. It is mostly in the northern countries that they become an item of great expense. In these countries, also, opium-eating, when carried to excess, becomes as ruinous to the purse as to the health.

In looking upon these estimates, we should not forget that stimulants and narcotics are a vast source of national income.

Furthermore, it must be remembered that they take the place of ordinary food, within certain limits, and thus, in that direction, enable us to economize while we are apparently so extravagant, and yet there can be no question that on the score of expenditures alone —waiving all considerations of health or morals—stimulants and narcotics are a serious evil to the poorer classes, especially of the United States and northern Europe. They serve to increase their poverty and to intensify its accompanying sorrows. They compel them to restrict themselves in the quality and quantity of bread and meat, and so they suffer all their lives long from insufficient nourishment.

How shall any individual determine whether he is benefited or injured by the stimulants and narcotics that he uses?

A friend of mine once expressed to me the wish that he had a little bell in his stomach which should strike as soon as he had eaten just enough.

Such ready-made contrivances for saving thought and care in matters of hygiene are not provided for us in this world. We are left to settle these hard and complex questions by patient and long-continued study of the history of races, nations and communities ; by close and discriminating observation of our personal experience from **day to** day, and from year to year, and only those **who think** that health and **life,** and the usefulness and happiness of ourselves **and of all who** may **succeed us are** of little worth, will hesitate to undertake **and pursue** this method. Mathematics are of but little **service in** solving this question. The estimates that have been made of the **number** of ounces of alcohol that can be daily taken, are of no practical value.

The following general considerations, together with **the** facts **already given, may be** of service.

Nervous temperaments cannot usually bear stimulants and narcotics of any kind, as well as the phlegmatic or the bilious temperaments.

Organizations **that are** generally susceptible to irregu **larities of diet, to changes of** the atmosphere, to the fatigue **of over** work, or to mental impressions of any kind, **are usually** susceptible **to** stimulants and narcotics.

Nervous **people,** therefore, need **to be** especially cau **tious in** forming habits of indulgence.

Those who live much in the open air, and labor with muscle as well as brain, can use them more freely than those who live mostly indoors, and are engaged in exclusively mental occupations.

Muscular exercise, to a certain extent, counteracts the evil effects of over **use** of these agents ; **but** in nervous organizations this counteracting power soon finds its limit.

According to Marcet, shopkeepers who live indoors are much more liable to chronic alcoholism than coal porters, who, although they drink more freely than shopkeepers, yet spend much of their time in the open air.

IMMEDIATE EFFECTS THE BEST GUIDE.

Although it is probable that stimulants and narcotics may exercise a slowly pernicious influence on the nervous system, without giving cause for suspicion by any immediate or temporary symptoms, yet, as a rule, there is little question that in nearly all such cases, close and intelligent observation can detect some evil effects after each act of indulgence.

So long as the immediate effects are good the permanent effects will probably be good ; so long as the immediate effects are evil, the permanent effects will be evil.

The great problem is to distinguish rightly between those effects that are good and those that are bad. The distinctions already drawn between *stimulation* and *narcosis* (page 6) will be of service in solving this problem. But there is always room for deception, especially for those who wish to be deceived. In our civilization continuous vigilance is the price of health, and those who think it too dear, have the choice of dispensing with the luxury.

I do not know, however, that this difficulty of determining what does and what does not injure us, applies very much more to negative than to positive food.

SPECIAL CONSIDERATIONS FOR AMERICANS.

Probably about half of the adult male Americans can smoke a cigar or two a day without apparent injury ;

but I am not one of the number. Of the other half, probably the majority would do as well, or better, without it. The number of those who can smoke a dozen cigars a day is so small in this climate, and with an American constitution, that no man has a right to try the experiment. It is like sitting all day with wet feet ; we may not take cold, but it is too risky.

Tobacco chewing and snuff taking, like dram drinking at public bars, will among cultivated people take care of themselves. It is hardly necessary to advise against them any more than it is necessary to advise against the habit of eating with our fingers, or swearing in drawing rooms. These habits disappear gradually before the rise of refinement, as marshy exhalations disappear before the rising sun.

The Americans are by no means the greatest smokers in the world. I think that nearly all the Europeans smoke more than we, and can bear it better. If the Germans are injured much by their almost constant smoking, it would be hard to tell just how, for both in muscle and in brain they are about the strongest in the world ; and dyspepsia is so rare among them, that they call it the " American disease." In Germany they have, as travelers know, cars for not smoking, just as we have smoking cars. The non-smokers are the exceptions.

The number of those in this country who find that they cannot use tobacco in any form without injury, appears to be continually increasing.

Many of my patients tell me that they have had to give it up ; that, without any doctor's aid, they knew that it was hurting them. I am inclined to think that at the present time, more among the higher classes of this country are physically injured by the slow and

silent effects of **tobacco, than by** alcoholic **liquors.**
I certainly should come to **that** opinion **if** I should
judge from **my own** patients.

In regard to the habitual use **of opium as a luxury, I**
can only say that, **for our climate and** constitution, I
would rather **risk my life by** jumping **off Niagara** Falls,
than by forming the habit of opium-**eating.**

And yet I know very well **that the writings** of De
Quincey are horrid exaggerations, or at least **have no**
general application ; and I believe **that there is a man**
in New York, over a century old, and in good health,
who, for a very long number of years, has eaten enor-
mous quantities of opium.

Coffee, like tobacco, is getting out of favor with **many**
of our Northern Americans, because they find that **they**
cannot use it. We bear none of these things **as our**
fathers did.

TEA USED TOO STRONG.

Tea is used too freely by many, and too strong by
nearly all. In China, where tea grows, and in Russia,
where it is used more freely than **in any** other country,
except **China, tea is** made very **weak,** being prepared not
by long steeping, **but by pouring boiling water on the**
leaves. **It benefits** more, and injures less, **for our cli-**
mate, and **for all cold climates, than any other form**
of stimulant. **Travelers in** Siberia find that tea **pre-**
pared in the Chinese **and** Russian fashion answers better than any other stimulant ; **it is there used in enor-**
mous quantities.

One more consideration I **will suggest** for the **en-**
lightment **of** conscience: it is, **that in this country**
it is possible **for large** classes **to work hard with the**

brain and attain a great longevity without using any stimulants and narcotics whatever, or only the milder forms.

Clergymen abstain more than any class, and although very nervous, they live longer than any other class in the country—longer even than farmers. Farmers as a class in these days are temperate, and they come next to clergymen in order of longevity. Then follow physicians, lawyers, merchants, the most abstemious classes we have in society. Abstinence, to say the least, allows of longevity.

ATTITUDE OF PHYSICIANS AND SCIENTISTS TOWARD THE TEMPERANCE REFORM.

Wonder has been felt that physicians and scientists, who ought to be, and are better qualified to form intelligent opinions on this subject than any other class, have so generally stood apart from this work. The reasons for this apparent indifference on the part of physicians and scientists are these :

1st. We did not know what stimulants and narcotics were or were not good for, hygienically or medically. It takes time to settle such questions, and long experimenting, and patient training of many minds. There has been no agreement among ourselves. If we should call up the fifty thousand or more physicians of this country, and ask them their opinion of each narcotic and stimulant in detail, we would be very likely to get fifty thousand or more different answers.

The number of *educated* physicians who, on hygienic grounds alone (without reference to the moral aspects of the case), enforce total abstinence from all alcoholic liquors, is probably very small : not more, I should

judge, than one in a hundred ; and many are in honest
doubt just what is the best course.   Naturally we have
wished to teach ourselves before we taught others.

Least of all can we be blamed for thus talking and
disputing.   When theologians cease their differences and
all warring creeds are merged in one ; when Re-
publicans and Democrats are all lost in one great
party who care not for spoils and love only liberty,
—then the world may pay its compliments to the
doctors and make them all agree.

ERRORS OF TEMPERANCE REFORMERS.

Another reason why scientific men in general, have
been cold toward temperance, is the extremely unscientific views and conduct of the temperance leaders and
orators.   In almost any daily or weekly paper that
is friendly to temperance, there are statements that
make the hair of a scientific man, or any one who
is accustomed to scientific modes of thought, stand on
end.   There are men of science whose writings are quoted,
but usually they are those whose views in the profession
have been exploded or are received only by a few ; and
it should be remembered that not every one who writes
on science is an authority in science, or regarded so by
his peers.   Here I may say that the enemies as well as
the friends of temperance have formed unwarranted
inferences from scientific writers.

Lyman Beecher is credited with the remark that
" When God has any great purpose to perform in this
world he makes some good man a little wrong headed
in the right direction."

In the Temperance Reform there have been many

good and great men ; there have been clergymen who represent the best brains and best culture of the land, and some of them have been not a little wrong headed.

If the time which has been spent in quarreling, in trying to set up patent methods of reform, had been concentrated in the effort to reduce the ignorance and extreme poverty of the land, to spread abroad sound knowledge on this subject, especially among the young, we should now be better off by far.

And yet the Temperance Reform is one of the most successful of efforts of modern or ancient times. We found intemperance fashionable : it has been made unfashionable. We found it in all grades of society : we have driven it, and are driving it from the college, the school, and the church, and the counting house, and are confining it in the natural home of all crime—the abodes of ignorance—and all this has been done without the sword, by moral and intellectual force alone.

Besides all this, physicians have been accused, and very unjustly, of causing much of the intemperance of the country. Cases now and then occur, where patients taking alcoholic liquors in disease, begin a habit which conquers them ; but I have never seen such a case. Prof. Austin Flint, one of our highest living authorities in medical science, and whose experience has been very great, tells me that he never knew a case. Recently, however, I have seen an instance reported in one of our medical journals. Whatever is given as medicine, oftentimes becomes repulsive.

Once I was told of a patient who was going down to the gutter and blaming his doctor for it ; but on searching the matter, I found that he had been a drunkard ten years before the doctor prescribed for him. There

are those in the profession who hold to the opposite
view that much of intemperance is caused directly
and indirectly by alcohol prescribing.

There is, I think, more truth in the charge that opium
eating is caused by opium prescribing ; although on
this subject I do not speak from personal observation,
but we will shower a great many pretty wreaths on the
man who will discover some remedy equally efficacious
as opium, and that leaves no bad effects.

There has been a disposition to criticise any one
who did not see eye to eye with the leaders of reform
on this matter. When scientific men, especially, speak
on the subject, a score of eager orators are watching
him, ready to pounce upon him if they think he blunders.
There are men whose views are worth hearing on this
theme, but who would rather, like Blondin, walk over
Niagara on a tight rope, than lecture on stimulants and
narcotics before a popular audience. A slip of the
tongue, a word that may be misunderstood, and away
they go down the torrent and over the rocks.

We have been afraid to let men think on this matter. A subject requiring more thought, more knowledge,
more wisdom, and more liberty of expressing opinions
in order to obtain this knowledge, than any other social
question of the world.; and yet we have cried out, "beware of the man who has any new facts or new thoughts."

I believe with Pascal that pure truth is not for us in
this world. I believe that error, like truth, has its mission : that falsehood may be one means of reform. I
believe that all the exaggerations and absurdities of the
temperance leaders in the past, have in some way
worked for good.

I believe with Herbert Spencer, that "Enthusiasm,

pushed even to fanaticism, is a useful motive power—perhaps an indispensable one."

Another reason why physicians have kept quiet is, that no one listens when they speak. In this land especially, science has few hearers. A funny story will bring a thousand where a world-touching fact, that has cost years of labor, brings one. Those who abuse science have the largest audiences.

It would have been better and wiser if scientific men had not waited for absolute and perfect knowledge on these themes, and had forgotten the bad treatment they had received, and joined the workers in the temperance army.

SHOULD WE ABSTAIN FOR THE SAKE OF EXAMPLE?

The question now arises whether it is not our duty in individual cases to forego the benefit and the pleasure that we may receive from any of these substances, and especially from alcoholic liquors of all kinds, and tobacco, for the sake of our friends, who are weak, our children, who ought not to use them, and the cause of temperance ; whether such self-denial will not add weight to our words ?

This is a question, not of science at all ; but of morals —of conscience.

In this, as in all similar questions of casuistry, we must work out our own salvation.

We must decide this question in the full light of all the knowledge we can command, just as we would decide what religious creed we will adopt ; just as we decide whether we will or will not attend the theatre, the opera, or the circus ; just as we decide whether we will dress in silks, and laces, and diamonds, while honest poverty is hungry and cold two blocks from our door.

The law cannot decide these questions for us ; and we cannot agree in our decisions, and it is not necessary or even desirable that we should. Just as no two faces are precisely alike, so no two brains are precisely alike, and so no two persons can think exactly alike. Our consciences, like the clocks of Charles V., will not tick together.

But however widely we may differ in questions of casuistry or in the local measures for the cure of intemperance, we can all agree in the constitutional treatment, which, after all, is the one great thing needful.

Every man who directly or indirectly helps to make less the ignorance of the poor, or of the rich ; who helps to solve the great labor reform problem, which is the right wing of temperance ; who helps to raise the race, or any one member of the race, one step higher, is a worker for temperance.

With all our vast immigration—every ship from Europe bringing a load of drunkards to our shores—with poverty and ignorance increasing, and the gulf between the higher and lower classes widening; with an increasing reverence for dollars above ideas, the outlook for immediate or absolute success is not flattering.

Just before us it is dark and lurid ; we cannot well see through to the light beyond.

Our consolation is that we work not for to-day, but for the ages to come.

The future of Temperance must inevitably share in the future of Civilization.

THE END,